Fundamentals of Soft Matter Science

Fundamentals of Soft Matter Science

Second Edition

Linda S. Hirst

CRC Press
Taylor & Francis Group
Boca Raton London New York

CRC Press is an imprint of the
Taylor & Francis Group, an **informa** business

CRC Press
Taylor & Francis Group
6000 Broken Sound Parkway NW, Suite 300
Boca Raton, FL 33487-2742

© 2020 by Taylor & Francis Group, LLC
CRC Press is an imprint of Taylor & Francis Group, an Informa business

No claim to original U.S. Government works

Printed on acid-free paper

International Standard Book Number-13: 978-1-138-72444-0 (Paperback)
978-1-138-72478-5 (Hardback)

Visit the Taylor & Francis Web site at
http://www.taylorandfrancis.com

and the CRC Press Web site at
http://www.crcpress.com

This book is dedicated to my parents, Bill and Sandra Matkin.

Contents

Preface

The goal of this book is to provide a comprehensive introduction to the science of soft materials. The study of different soft matter systems has a long history but, until relatively recently, the subject was not consolidated into a distinct field. The idea that common physical principles underlie many of the once-separate soft matter topics was pioneered by 1991 Nobel Laureate Pierre-Gilles de Gennes. His theoretical work inspired a new way to approach and define soft matter, inspiring huge interest and growth in this fascinating subject. In addition, the parallel rapid growth of biophysics and polymer science over the past few decades has produced an explosion of focused soft matter research groups worldwide. Nowadays, soft materials are at the forefront of many contemporary technologies such as novel flexible displays, tissue engineering scaffolds, lab-on-a-chip biotechnology applications, and cutting-edge food science.

Recently it has become increasingly common for professors to include soft matter topics in their classes, or to offer complete courses, either at the graduate or undergraduate level. This development is extremely positive for the field of soft materials. Consequently, soft materials such as polymers, rubbers, gels, colloids, and liquid crystals touch every aspect of our lives and should be a standard part of the undergraduate curriculum. How can students understand the properties of everyday materials without an introduction to the fundamental science behind molecular self-assembly and the physics of soft materials?

Therefore, my goal was to write a readable book designed to be understandable at the undergraduate level for students with a background in introductory college-level physical and chemical sciences. The book is designed to be used as a textbook introduction for mid- to upper-level undergraduates and also to act as a useful resource for graduate students or more senior scientists of all levels new to the field of soft matter.

With these goals in mind, the book puts an emphasis on conceptual understanding and tends to avoid a lot of mathematical derivations. Lower division/introductory college course knowledge in physics, chemistry, and math is assumed throughout, with some refreshers and the appendices fill in some mathematical details. The aim is not to provide a rigorous theoretical description of any particular subject because this work has already been done very well by many authors at a graduate level in the specialized books among those referenced at the end of each chapter.

I have broadly defined the materials into several classifications. These classifications are by no means distinct, however, and there are many soft materials around that can span over a couple of chapters in this book. There are also many crossovers between subject areas in terms of theoretical descriptions and experimental techniques. In fact, one of the most fascinating aspects of soft matter science lies in the many conceptual connections that can be drawn between different materials. For example, much of the chapter on surfactants could potentially be classified as part of the liquid crystals chapter, and certain topics in the biomaterials chapter could equally find a home as part of the surfactant or polymer chapters. One important concept to understand from this book is the universality of the physics we use to describe soft materials. Although scientists and students may identify with one or more of the main topics presented here, there is much overlap and flexibility in how materials can be described and classified.

Each chapter in the book is dedicated to a different group of soft materials. Through the book we will look at liquid crystals, surfactants, polymers, colloids, and selections of soft biomaterials. For each subject, I discuss the essential concepts of the subfield: material structures and physical characteristics, some simple theoretical ideas, and important experimental methods. Italicized terms can be found in the glossary for quick reference. As there is a natural overlap in experimental techniques between the chapters I have tried to find the most comfortable home for each technique.

At the end of each chapter, a further reading section will guide the reader toward more in-depth future study, including some of the detailed graduate-level texts currently available. My hope is that this book will serve as a springboard for students of all levels to initiate and expand their knowledge of soft matter, and perhaps even consider a future career or new research program specializing in some aspect of this fascinating subject.

Acknowledgments

I would like to give special thanks to the many colleagues and friends who contributed their time and expertise to help with the writing of this book. Enormous thanks to all my students who gave me feedback on the materials and helped me to improve the text for this new edition. Finally, I want to thank my parents, Sandra and Bill Matkin, to whom this book is dedicated and my husband, Trevor—not only a talented photographer and illustrator, but the love of my life and an amazing source of support and inspiration.

About the Author

Linda S. Hirst is Professor of Physics at the University of California, Merced where she has been a faculty member since 2008. Born in Liverpool in the UK, she obtained her B.Sc. and PhD in Physics from the University of Manchester and spent time as a postdoctoral researcher in the Department of Materials Science and Engineering at the University of California, Santa Barbara. Before joining the University of California, Merced, Professor Hirst spent 3 years on the faculty in the Physics Department at Florida State University. Professor Hirst's research interests cover a variety of topics in soft-condensed matter physics, with particular interests in soft biomaterials and liquid crystals. She has a strong interest in promoting the study of soft matter and its importance to students, scientists, and the public.

About the Illustrator

Trevor Hirst read Law at the Victoria University of Manchester, went on to train as a solicitor (attorney), and was admitted to the Roll of Solicitors of the Supreme Court of England and Wales in 1999, practicing civil litigation in Altrincham, Cheshire, UK. He and his wife, Linda, emigrated to California in early 2002, whereupon Trevor changed careers from litigation to education and began to follow more creative pursuits such as photography and illustration.

He now works for the University of California, Merced, as the executive director of the Health Sciences Research Institute, where his primary responsibilities are management of institutional operations, research development, and external relations.

Introduction

1.1 WHAT IS SOFT MATTER?

Soft matter science is the study of materials that are physically "soft," and this book is an introduction to the science behind these materials. But, what does it mean to be soft? The concept of a soft material can be applied to many different substances, ranging from relatively hard plastics and rubbers all the way to a variety of apparently liquid-like materials, such as polymer solutions or colloidal suspensions. A more precise definition of a soft material is therefore complicated somewhat by the huge range of physical properties observed in these systems. Pierre-Gilles De Gennes, the Nobel Prize-winning theoretical physicist, characterized soft matter as a class of materials that give a "large response to small perturbations."[1] This idea provides us with a clear and straightforward description. Any material that deforms easily under an external stimulus (e.g., mechanical forces, electric or magnetic fields, etc.) is "soft."

Another helpful concept that can help us to understand and identify soft matter is linked to thermal energy and intermolecular interactions. Soft materials are held together by weak intermolecular interactions and have a structure that can be significantly affected by energies close to the thermal energy $k_B T$, where k_B is the Boltzmann constant (1.38×10^{-23} m^2 kg/s^2 K), and T is the material's temperature measured in Kelvin. At room temperature, $k_B T$ has a value of about 26 me V (or 4.2×10^{-21} Joules)—much smaller than the energies that hold atoms together in a crystalline solid. Therefore, one of the most important characteristics of a soft material is that its structure can be easily altered at relatively low temperatures.

A third characteristic of soft materials is that they often have some degree of self-assembled non-crystalline structure. The basic component units of the material are arranged in a non-random fashion, although this ordering is typically short-range in comparison to the long-range regular atomic structure of a solid crystalline material. In terms of their structure, soft materials exist in a relatively disordered state with a molecular arrangement somewhere between a crystalline solid and a conventional liquid.

When we study the structure of materials, we usually first learn about the three basic phases of matter: solids, liquids, and gases. While these definitions provide an essential scientific foundation, they fail to describe adequately the majority of familiar materials we see around us and interact with in our everyday lives, including plastics, gels, rubbers, soaps and other detergent products, paints, most foods, and most of the human body (Figure 1.1). The aim of this book is to introduce a new flavor of materials science focused on these soft materials. Through descriptions of the different classifications of soft materials and their structures, we will also provide an overview of some common experimental techniques and basic theoretical ideas.

There are several physical concepts that will be important to your understanding of the material in this book:

1. Basic thermal physics and phase transitions
2. Intermolecular forces
3. Self-assembly
4. Mechanical properties of materials

FIGURE 1.1   Examples of some of the different soft materials explored in this book.

As these ideas are so important and commonly encountered throughout the field of soft matter science, we will review them over the first two chapters of this book. In this first introductory chapter, we will spend some time on important concepts from thermal physics, including temperature, phase transitions, and phase diagrams.

1.2   BASIC THERMAL PHYSICS

Temperature is perhaps the most important experimental parameter for soft matter science because the structures of soft materials are so sensitive to energy changes on the order of $k_B T$. (4.2×10^{-21} J or 0.026 eV at 300 K). Because soft materials can be so sensitive to small temperature changes, random thermal molecular motions act to define phase behavior and structure. The concepts of thermal equilibrium, phase behavior, and statistical physics are central to both a basic and a more advanced understanding of this field, so we must begin there.

1.2.1 EQUILIBRIUM

In general, when two objects are in good thermal contact with each other (i.e., heat can transfer readily between them), they will eventually come to same temperature. At that point, the objects have reached a state we call *thermal equilibrium*. The timescale over which this occurs is known as the *relaxation time*. For example, if a metal cube is heated to 90°C and placed into a glass of water at 20°C, the two materials will come to thermal equilibrium. The temperature of the water will increase and the temperature of the metal cube will decrease until a common temperature is reached somewhere between 90°C and 20°C. The fact that systems spontaneously relax to equilibrium is one of the most important concepts in thermodynamics. Of the many different kinds of soft materials you will come across, most are usually found in their equilibrium state. Once in equilibrium, a system will tend to stay that way.

A similar concept can be applied to mixing liquids, but the idea of equilibrium between two or more populations of molecules in a liquid requires us to go one step further. In addition to their thermal equilibration, the molecules in a two-component liquid can diffuse around each other, plus they may also interact (by attraction or repulsion). Such interactions can result in the spontaneous formation of stable structures as the different molecules arrange themselves favorably. Take for example, the addition of cold milk to a hot cup of tea. In the cup, the milk is in good thermal contact with the tea, and the two liquids come to thermal equilibrium. The two liquids will also mix gradually (by diffusion or stirring) until both tea and milk are uniformly distributed. Milk, however, is not a simple solution, but actually an aqueous suspension of tiny fat droplets, surrounded by stabilizing proteins. In equilibrium, the fat molecules do not disperse throughout the tea (fat molecules should separate from water), instead, they tend to stay with the proteins, and under the right conditions these stabilized droplets suspend with a uniform distribution throughout the tea. This simple example highlights the fact that there are often fascinating structures underlying apparently simple systems. Throughout this book you will read about many examples of complicated mixed systems, including detergent molecules in water, polymer gels containing a solvent, and particles in suspension in a liquid phase. In most cases, the different populations of molecules are in thermal and chemical equilibrium, so when left for a long time at the same temperature and pressure, the material will not vary in macroscopic structure. Consider that the equilibrium state of a system may not be the most uniform, but in fact can have a complex microstructure.

In addition to materials in equilibrium, there are many examples of systems that, while maintaining a steady state, are in fact, not in equilibrium. Some materials exist in what we call a metastable state, i.e., they are trapped out of equilibrium in a local energy well. Instead of reaching their lowest energy state, they have stabilized in an intermediate structure. There are many different examples of this behavior in polymer or colloidal systems where the constituent particles or molecules can become entangled or jammed. Active systems represent another class of out-of-equilibrium materials. In contrast to equilibrium systems, active fluids and systems of active particles consume energy from the environment. They often exhibit self-generated internal flows and behaviors such as swarming and clustering. Examples of active matter can be seen in self-propelled colloids and ATP-driven biological assembles. We will take a look at an example of an active phase later on in Chapter 7.

1.2.2 PHASE TRANSITIONS

A phase is an equilibrium state of matter with a distinct structure that does not vary in time. This structure can be described by the average molecular arrangement of the component molecules. For example, ice and liquid water are two different phases of the same material, but oil and water can be two different compositional phases in a mixture. In a two-phase system, there will be a clear boundary between the two states of matter. We can map the presence of different phases in a pure or mixed material using a phase diagram. The phase diagram shows the equilibrium conditions under which certain phases occur (i.e., temperature, pressure, composition, etc.); we revisit this idea in more detail in Section 1.6.

We are most familiar with the three standard phases of matter: the solid, liquid, and gaseous states (Figure 1.2); however, the idea of a phase can be much broader. In fact, there are many different states of matter that can be defined inbetween the solid and gas phases, and we will learn about a number of them in this book. Soft materials often exhibit complex phase behavior with many different phases possible under different conditions (i.e., temperature, composition, pressure, etc.). Understanding the structure and properties of these phases and how we can move between them is at the heart of soft matter science.

At a phase transition, a material will suddenly transform from one state to another; for example, a melting crystalline solid will go from a regular lattice structure to a disordered fluid. This transformation may have an energy associated with it, and we can define the *heat of transformation Q*:

$$Q = Lm \tag{1.1}$$

where L is a constant associated with the particular transition and material, and m is the mass of the material. For example, the latent heat of vaporization L_v for water is 2260 kJ/kg at a temperature of 100°C. This is the amount

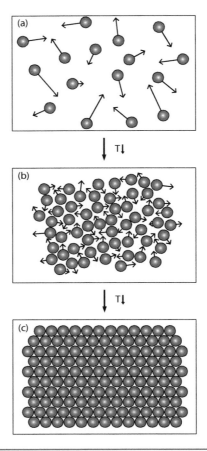

FIGURE 1.2 The three most well-known phases of matter—(a) gas, (b) liquid, and (c) solid—shown as spheres in a box. The material undergoes two phase changes as the temperature (T) is decreased. Each sphere represents an individual molecule or atom in the system, and the arrows indicate molecular velocity vectors. In the case of the solid, atoms in the lattice will vibrate, but they cannot move independently from their lattice positions, so no arrows are shown.

of heat measured in kilojoules released from vaporizing 1 kg of liquid water. The latent heat of fusion L_f is the amount of heat needed to melt 1 kg of ice. Latent heat represents the stored energy in the lower-temperature phase released on heating the material to a temperature above the phase transition point; Table 1.1 shows some values for common materials. At a phase transition, this latent heat is equal to the change in the *enthalpy* of the system (ΔH).

Phase transitions in a material can be classified as *first order* and continuous (second order). At a first-order phase transition, we will observe

TABLE 1.1
Latent Heat of Fusion L_f and Vaporization L_v for Some Common Materials

Material	L_f (kJ/kg)	L_v (kJ/kg)
Water	334	2,260
Ethanol	109	879
Gold	63.7	1,645
Carbon dioxide	571	205
Copper	209	4,730
Mercury	11.4	295
Silicon	1790	12,800

Note: Notice how these values compare to kT at room temperature.

a discontinuity of some physical property representative of the degree of order in the system. For example, this could be material density or a calculated measure of order in the material (i.e., an order parameter). If we measure this parameter across the phase boundary, there will be a step, or discontinuity, at the transition point. An example of a typical first-order phase transition is ice melting. At the transition point, the density of the material abruptly changes as we go from ice to liquid water. First-order transitions also have a measurable latent heat. Some phase transitions can be described as "weakly first order." In this case, the enthalpy change associated with that transition is very small. This is often true of phase transitions in soft matter systems. As a result, the enthalpy change may be difficult to measure, making the phase change difficult to detect by thermal properties alone.

Second-order phase transitions do not have an associated latent heat, and their density along with other important order parameters varies continuously across the phase boundary.

1.2.3 SOLIDS, LIQUIDS, AND GASES

When discussing how materials behave in response to a temperature change, we should be familiar with the concept of their "phase." Usually, this means whether the material is a solid, a liquid, or a gas, and a brief review of these basic states will help us to understand many concepts in soft materials. A gas model, such as the ideal gas can be used to describe dilute systems with minimal inter-particle interactions. However, in the denser liquid state, inter-particle interactions become important, and we will describe their effects in the following sections.

(a)   (b)

FIGURE 1.3   Examples of different solids, (a) amethyst, a crystalline solid and (b) glassy amber, an amorphous solid.

Solids can be either crystalline or amorphous (Figure 1.3), but in both cases the atoms that form the solid are very close together. The material is difficult to deform and has a well-defined shape. On heating, some increase in the volume may occur, but this change is small in comparison to the overall size of the object. Crystalline solids are highly regular in structure, have long-range order, and can be characterized by a unit cell, the smallest possible repeat unit in the crystal. Metals, salts, and precious stones are examples of crystalline solids.

There are also many amorphous solids around us, in fact probably more than there are crystalline solids. Most of the amorphous solids that you will see are plastics and glasses. These materials do not have a long-range regular structure: The molecules that make them up are organized randomly, "frozen" into that arrangement as the material was cooled during the manufacturing process.

In a liquid, the molecules are not frozen into a set position, but are free to move throughout the material. Liquids are relatively dense and incompressible compared to the gas phase; therefore, intermolecular forces play an important role in the properties of the liquid. Intermolecular forces are extremely important in soft materials and are discussed later in this chapter. The liquid state is the most relevant of the three basic phases of matter, and many of the materials and phases we will discuss have fluid-like properties. Another definition of soft materials is that they are "structured fluids": The material may flow or exhibit diffusion properties in common with simple liquids, but there is some degree of ordering that can be measured in its molecular arrangement.

1.2.4   THE IDEAL GAS

In the gaseous phase, we can usually think about the molecules being on average very far apart from each other and moving very quickly. The behavior of a gas can be described using the *ideal gas model*. In an ideal gas, we assume that:

1. The molecules (or atoms) in the gas are very small point masses and therefore do not take up any volume
2. Intermolecular forces are negligible

This successful model for dilute gases works very well for systems that are far from a phase transition and not under high pressure. That is, for the model to work, we make sure that the molecules are not, on average, sufficiently close to each other to experience any significant intermolecular forces, or that their physical size needs to be taken into account when describing their behavior.

The speeds of the molecules in an ideal gas are distributed over a wide range of values. At a given instant in time, if you were to take a snapshot of the gas, measure the instantaneous speed of every molecule, and plot the results as a histogram, you would obtain a Maxwell-Boltzmann distribution of speeds. Some examples of this distribution function at different temperatures are shown in Figure 1.4; higher temperatures lead to a greater proportion of molecules moving with high velocities, although there are always significant numbers of slow-moving molecules in any system as a result of the random collisions they undergo.

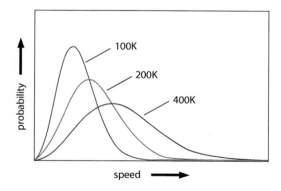

FIGURE 1.4   The Maxwell-Boltzmann distribution for different temperature gases. At three different temperatures, the plot represents the probability of a gas molecule in the system having a particular speed.

The average kinetic energy *KE* of a molecule in a monoatomic ideal gas is given by:

$$KE_{av} = \frac{3}{2}k_B T \tag{1.2}$$

½ *KT* for each translational degree of freedom (*x*, *y*, and *z*).
So for a single molecule of mass *m* we can use:

$$\frac{1}{2}m\bar{v}^2 = \frac{3}{2}k_B T \tag{1.3}$$

and thus obtain an equation for the root mean square (rms) velocity of a molecule in the gas:

$$v_{rms} = \sqrt{\frac{3k_B T}{m}} \tag{1.4}$$

This value for the rms velocity falls close to the peak of the curve in Figure 1.4, although the value is always slightly shifted to the right of the peak due to the asymmetry of the curve. The exact shape of this curve is given by the following equation:

$$D(v) = \left(\frac{m}{2\pi k_B T}\right)^{3/2} 4\pi v^2 e^{-mv^2/2k_B T} \tag{1.5}$$

where *D*(*v*) is the probability distribution for molecular speed in the gas. At high temperatures, this curve dies away exponentially. A simple derivation for this formula can be found in the popular thermodynamics textbook, *An Introduction to Thermal Physics* by Daniel V. Schroder.[2]

The macroscopic properties of the ideal gas can be described very well in terms of pressure *P*, temperature *T*, and volume *V* using the *ideal gas equation*.

$$PV = nRT \tag{1.6}$$

This useful and well-known equation relates these parameters using the ideal gas constant (*R* = 8.314 J/K · mol) for *n* moles of gas.

Although the ideal gas is not a good model for soft matter—which is much more condensed than a gas of course—it is an important to think about, as it helps us to gain intuition when thinking about molecular motions in fluid systems and intermolecular interactions. Soft materials are usually condensed fluid phases, in which excluded volume and intermolecular interactions play a very important role. They are also very temperature-sensitive and thermal molecular motions contribute significantly to their macroscopic properties.

1.3 PHASE DIAGRAMS

Materials can change their phase as a function of a variety of different thermodynamic variables, such as temperature, pressure, volume, or concentration in a mixture. A diagram of this phase behavior, or "phase diagram," can be constructed by plotting the parameter ranges over which different phases occur. The simplest form of the phase diagram plots the position of the phases as a function of just two parameters, such as the pressure/temperature phase diagram for water shown in Figure 1.5; however, it is also common to see graphical representations of three or even four different parameters on the same diagram. We can refer to the boundaries of phase existence in terms of thermodynamic parameters as the phase's position in "phase space." If a material system exhibits several different phases, then the phase diagram can become quite complicated.

Looking at the phase diagram for water, we can identify some interesting points. The solid red lines represent boundaries between the different phases, and as a reference point, the boiling point of water at atmospheric pressure (101,330 Pa or 1 atm at sea level), is indicated. As we can see from the diagram in Figure 1.5, the temperature at which water boils can vary significantly with a change in pressure; even relatively small changes in pressure can give a noticeable difference, and this effect can be easily tested in real life. At high

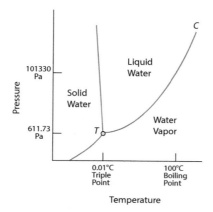

FIGURE 1.5 The pressure/temperature phase diagram for water indicating the triple point at point *T* and a critical point at *C*. The phase boundaries are indicated by the solid red lines.

altitude, food takes longer to cook because water boils at a lower temperature at lower pressures; in fact, in the "mile-high" city of Denver, Colorado, the air pressure is so reduced that water boils at close to 95°C.

Point T in Figure 1.5 marks the meeting point of the three different phase boundaries and is known as the *triple point*. The triple point occurs at a unique temperature, volume, and pressure for a material, and at this point all three phases (solid, liquid, and gas) may coexist. The triple point of water is particularly important as it provides the definition of the Kelvin (K), the SI unit of thermodynamic temperature. In 1954, the General Conference on Weights and Measures (*Conférence Générale des Poids et Mesures*, GCPM), an international organization responsible for the SI system of measures, defined the Kelvin unit as "the fraction 1/273.16 of the thermodynamic temperature of the triple point of water."[3]

At point C, we can note another interesting feature of the diagram. The line that defines the transition between liquid and gas is discontinuous and ends at this point. This is a *critical point*. Below the critical point, the phase transition is discontinuous, with an associated latent heat (first order), but above this line there is no defined phase transition from liquid to gas, and the density of the material varies continuously.

THE CLAUSIUS-CLAPEYRON EQUATION

The lines on a phase diagram represent the boundaries between thermodynamically stable phases, and close to these lines it takes just a very small change in temperature or pressure (some other parameter) to go from one phase to another.

Let us consider the well-known Pressure-Temperature phase diagram for water shown in Figure 1.5. Water in the liquid phase will have a Gibbs free energy G_l, and in the gas phase, a Gibbs free energy of G_g.

If we define a point that lies on the phase boundary between the liquid and gas phases, then at that point,

$$G_l = G_g \qquad (1.7)$$

This is true at any point on the phase boundary line. To move a small distance along the line, we change the temperature by dT and the pressure by dP; thus,

$$dG_l = dG_g \qquad (1.8)$$

Now, the Gibbs free energy is defined as

$$G = U - TS + PV \qquad (1.9)$$

(Continued)

THE CLAUSIUS-CLAPEYRON EQUATION (*Continued*)

Where S represents entropy using the first law of thermodynamics, $U = Q - W$ at our phase boundary, where U is the internal energy of the materials, Q represents heat flow and W is work we can write (see if you can make this step):

$$-S_l dT + V_l dP = -S_g dT + V_g dP \qquad (1.10)$$

This equation can then be rearranged to:

$$\frac{dP}{dT} = \frac{S_g - S_l}{V_g - V_l} \qquad (1.11)$$

Since we know that the heat required for the transition $\Delta Q = T \Delta S$ is equal to the latent heat L, this equation is rewritten as:

$$\frac{dP}{dT} = \frac{L}{T \Delta V_l} \qquad (1.12)$$

This result is the *Clausius-Clapeyron equation* and describes the shapes of the phase boundary lines on a pressure temperature (*PT*) phase diagram. In particular, this expression is useful for calculating the vapor pressures of different materials (for either evaporation or sublimation).

1.4 DIFFUSION AND RANDOM WALKS

Add a few drops of ink to a glass of water without stirring and you soon observe how the ink gradually spreads through the water, moving from areas of high concentration to areas of lower concentration. This mass transport process can be summarized by Fick's law:

$$J = -D \frac{dc}{dx} \qquad (1.13)$$

where J is the molecule flux (or number of molecules passing through a unit area per unit time). D is a constant, the diffusion coefficient, and c is the molecular number density. Therefore, dc/dx represents a molecular concentration gradient.

Diffusion is quite well known in solutions and gases, but also takes place in soft materials, for example, colloidal particles suspended in a liquid. Large particles, when placed in a fluid-like medium, will diffuse through the solution as a result of random thermal motions by the liquid molecules. The larger particles dispersed in the medium are subject to a phenomenon known as *Brownian motion*. We alluded to this phenomenon when discussing the thermal motion of particles. In a system comparable to the ideal gas, all particles in a system at a temperature T move randomly with a distribution of speeds. The speed distribution is represented by the Maxwell-Boltzmann distribution (Figure 1.4), and the average speed is related to the particle mass m. When a large particle or molecule is placed into a fluid comprised of smaller particles, many collisions between the different-size particles take place. All of the particles in the system are subject to random forces as a result of these collisions. The smaller particles move around rapidly and collide often with the larger particles, and this random bombardment results in a net force on the larger particle that will vary in direction with time. The overall average net force on a particle will be zero measured over a long time; however, at any specific time the random nature of the collisions will cause the particle to experience a fluctuating force. The large particle will accelerate in response to that changing force and move through the medium with a Brownian motion. Occasionally, the random force can be large, resulting in a significant jump in the particle's trajectory.

This kind of motion was first studied by a Scottish scientist, Robert Brown, in 1828. Brown was a botanist and observed the phenomenon as he looked at pollen grains on the surface of water. He noticed a random spontaneous motion of the grains and investigated the effect with a variety of small particles. It was not until the later nineteenth century, however, that this phenomenon was related to research being carried out on the kinetic theory of gases and the ideal gas. In fact, it was Einstein who published the definitive paper on the link between the statistical movement of particles and the macroscopic Brownian motion effect.[4] Interestingly, the idea was not generally accepted at the time because concepts of atoms and molecules were still in their infancy.

The motion of the Brownian particle can be described as a *random walk* because the particle appears to move disjointedly in random directions (Figure 1.6). An estimate for the 3D diffusion of a particle in solution can therefore be made using the following equation:

$$\lambda = (6Dt)^{1/2} \tag{1.14}$$

where λ represents the characteristic displacement of the particle at time t, and D is the diffusion coefficient. λ gives a measure of the size of the space

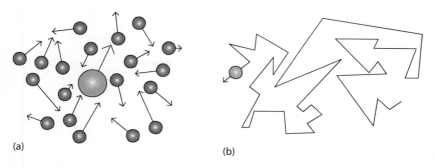

FIGURE 1.6   (a) Small particles bombard a larger (green) particle randomly, giving it a net motion. (b) If we watch the motion of the green particle over time, it can be described as taking a random walk.

explored by the particle in its random path after a certain time, or the average size of a random walk over time, t. This formula can be used to describe the diffusion of particles or molecules in a fluid over time. In real systems, the shape of the large particle is a factor in the diffusion rate, and at the length scales applicable to some soft materials, viscous drag on the particle becomes important.

The *Stokes-Einstein equation* that follows provides us with the diffusion coefficient for a spherical particle suspended in a liquid by taking into account the drag force on the particle:

$$D = \frac{k_B T}{6\pi\eta r} \tag{1.15}$$

Here, η is the viscosity of the medium, and r is the effective radius of the particle or molecule. As we will see in Chapter 6, the Brownian motion described here is applicable not only to spherical colloidal particles in solution, but also to the configuration of a polymer chain. These arguments relate the diffusion process to the kinetics of an ideal gas, so do not take into account intermolecular interactions. Additional attractive or repulsive forces within a fluid can have a significant effect on diffusion rates in soft materials.

FICK'S FIRST LAW AND THE DIFFUSION CONSTANT

Fick's first law describes the diffusion rate of molecules in a concentration gradient.

Let us define a plane of area, A, in a concentration gradient. Particles are more concentrated on one side of the plane. The particles move randomly and can pass through the plane from x to $x + \Delta x$. And back in 1D. The number of particles on the left side of the plane is $N(x)$ and on the right, $N(x + \Delta x)$.

Since the particles move randomly back and forth on both sides, there is a 50% probability that a particle on either side will cross the volume in time, τ, therefore the net flow of particles to the right is equal to,

$$\Delta N = -\frac{1}{2}\left[N\left(x+\Delta x\right)-N\left(x\right)\right] \tag{1.16}$$

This gives us a flux, J of molecules crossing the plane in time τ as,

$$J = -\frac{1}{2}\frac{\left[N\left(x+\Delta x\right)-N\left(x\right)\right]}{\Delta \tau} \tag{1.17}$$

If we let the particle concentration, $c\left(x\right)=\dfrac{N\left(x\right)}{A\Delta x}$, then we can rewrite as,

$$J = -\frac{\Delta x^2}{2\tau}\frac{\left[c\left(x+\Delta x\right)-c\left(x\right)\right]}{\Delta x} \tag{1.18}$$

Then, by defining, D, the diffusion constant as,

$$D = \frac{\Delta x^2}{2\tau} \tag{1.19}$$

We obtain Fick's first law,

$$J = -D\frac{dc\left(x\right)}{dx} \tag{1.20}$$

and see that the diffusion rate is proportional to the concentration gradient and the diffusion constant. Incidentally, then we can write the useful relation for 1D,

$$\Delta x = \sqrt{2D\tau} \tag{1.21}$$

where Δx represents a characteristic diffusion length over which a particle is expected to travel in time, τ. This can be expanded to two- and three-dimensional diffusion, adding the additional degrees of freedom and defining λ as the characteristic diffusion length, so $\lambda_{2D} = \sqrt{4D\tau}$ and $\lambda_{3D} = \sqrt{6D\tau}$.

QUESTIONS

Soft Materials and their Characteristics

1. Soft materials can be described as having a "large response function"; they deform readily under external stimuli. Look around you and identify some soft materials. How many of the materials around you are crystalline solids?
2. Give some examples of non-Newtonian fluids in everyday life. Would you consider these materials "soft matter"? Explain your answer.

Review of Thermal Physics

3. Which of these materials can be considered an equilibrium phase of matter? (a) a pile of sand, (b) a diamond, (c) milk, and (d) a stretched elastic band.
4. What is the physical significance of the negative slope of the ice/liquid water phase boundary on the water pressure/temperature phase diagram?
5. The pressure/temperature phase diagram for water shows that increasing pressure can melt ice. A commonly stated example of this effect is that the pressure on the blade of an ice skate melts a layer of ice beneath the skate, allowing the skater to glide smoothly. Do you think this idea is true? Use the Clausius-Clapeyron equation to investigate. What other factors could contribute to smooth skating?
6. A pollen grain on the surface of a lake appears to move around in a random, jerky motion. Can you explain this observed phenomenon to a high school student?
7. A droplet containing a concentrated solution of protein molecules is deposited at one end of a narrow microfluidic channel, 3 mm long, containing water. Assuming that the sole mechanism for molecular transport is diffusion, calculate the average time it will take for the proteins to reach the other end of the channel. You can assume that the protein molecules are 5 nm in diameter to estimate the diffusion constant, D.
8. Fick's 1st law gives us the rate of diffusion in a steady state system with a fixed concentration gradient, but does not take into account the time dependent variation in that concentration gradient. Fick's 2nd law takes the variation of concentration over time into account. Can you derive this equation using a similar approach to that used to derive Fick's first law?

$$\frac{dc}{dt} = D\frac{d^2c}{dx^2}$$

9. In an ideal gas, the distribution of molecular speeds is given by the Maxwell distribution function,

$$D(v) = \left(\frac{m}{2\pi k_B T}\right)^{3/2} 4\pi v^2 e^{-mv^2/2k_B T}$$

where m is the molecular mass, T is the temperature, k is the Boltzmann constant, and v is the molecular speed. Use this equation to derive a simple expression for the most likely speed in the gas. Why does this differ from the rms speed?

10. The van der Waals equation of state,

$$\left(P + \frac{aN^2}{V^2}\right)(V - nb) = nRT,$$

relaxes the assumptions of the ideal gas model by allowing for particle volume and inter-particle interactions. In this equation, P is the pressure, V is the volume, n is the number of moles, and R is the ideal gas constant. Two coefficients are introduced: the Van der Waals coefficients a and b. Constant a provides a correction for intermolecular forces and b corrects for finite molecular size (i.e., two particles cannot be in the same place at the same time). Notice how the equation reverts to the ideal gas equation when these constants are zero.

In an aqueous suspension of 1 μm radius particles, there are 10^8 particles/cm^3. Calculate the volume fraction occupied by the particles (the excluded volume). Compare your answer using He atoms and 10 μm radius particles. How does excluded volume fraction scale with particle radius?

11. Using the Clausius-Clapeyron equation, estimate the boiling point of water on top of Mount Everest.

REFERENCES

1. P.G. de Gennes, Soft matter: Birth and growth of concepts. In *Twentieth Century Physics*. L.M. Brown, A. Pais, and B. Pippard (Eds.). Boca Raton, FL: CRC Press, Chapter 21 (1995).
2. D.V. Schroeder, *An Introduction to Thermal Physics*. Boston, MA: Addison Wesley Longman (2000).
3. *The International System of Units*, 8th ed. Sèvres, France: Bureau International des Poids et Measures (2006).
4. A. Einstein, Theoretische Bemerkungen über die Brownsche Bewegung. *Z. Elektrochem.* 13, 41–42 (1907).

FURTHER READING

I.W. Hamley, *Introduction to Soft Matter: Polymers, Colloids, Amphiphiles and Liquid Crystals*–Revised Edition. New York: Wiley (2008).

R.A.L. Jones, *Soft Condensed Matter*. Oxford Master Series in Condensed Matter Physics, Vol. 6. Oxford, UK: Oxford University Press (2002).

M. Kleman and O.D. Lavrentovich, *Soft Matter Physics: An Introduction*. New York: Springer (2001).

D. Tabor, *Gases, Liquids and Solids, and Other States of Matter*, 3rd ed. Cambridge, UK: Cambridge University Press (1991).

Self-assembly and Structure in Soft Matter

2.1   SELF-ASSEMBLY

The term *self-assembly* has become ubiquitous in materials science over the past few decades, particularly in the field of soft matter and in related fields at the convergence of soft and hard materials. It is important for us to understand this concept, as it will be used frequently throughout the book. Self-assembly can be described as spontaneous molecular ordering resulting from the balance between entropic and intermolecular forces in a material. A self-assembled system or state is one that forms without external mechanical manipulation of the components. Instead, the elements of the material (molecules, particles, etc.) are subject to forces between these elements and thereby spontaneously adopt a particular configuration, or structure by coming to either an equilibrium or locally stable state.

Visualize a system composed of particles moving freely in a fluid at room temperature. The particles are attracted to each other; but, Brownian motion at room temperature keeps the particles constantly moving randomly. Here, we can see that there is competition between the attractive interaction—driving particle clustering, and the disordering effects of the thermal environment. This delicate balance determines the equilibrium structure of the material at a particular temperature. If the temperature is too high, thermal motions will prevent the particles from condensing into a solid structure of clustered particles; however, if the temperature is low, thermal motions will not be large enough to overcome the attractive potential, and the particles will stick together. This simple idea can be expanded to systems with many complicated interparticle interaction forces. The interplay between intermolecular forces (attractive or repulsive), and thermal motion drives the self-assembly of an enormous range of soft structures and phases, many of which are described in this book.

Self-assembly into a soft phase is most likely to occur if the material is in a fluid-like or semifluid-like state because for ordering to arise out of random fluctuations, the molecules need to be able to move around and try out different arrangements. It is statistically possible for the constituent molecules in a material to have any possible configuration, however unlikely, but the final structure is determined by the most likely and therefore energetically favorable molecular arrangement. Figure 2.1 shows two interesting examples of self-assembly in action: the spontaneous folding of a complex protein structure and the formation of a liquid crystal phase.

The term *self-assembly* is also often applied to the deposition of thin films on surfaces (for example, self-assembled monolayers, [SAMs]) in the field of nanoscience. The formation of self-assembled films can involve a variety of interactions, such as electrostatics, van der Waals forces, or even a chemical bond. Such non-crystalline molecular films should also fall

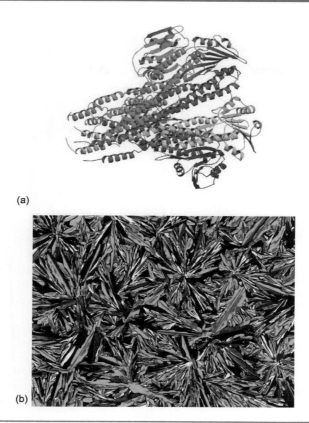

(a)

(b)

FIGURE 2.1 Two examples of self-assembly in action: (a) A ribbon diagram of the magnesium membrane transporter protein CorA (rendered using Pymol from Protein Data Bank file 2IUB). Proteins composed of a long amino acid chain assemble into a highly specific shape through chain–chain interactions. (b) Polarized microscopy image of the liquid crystal B1 phase in which banana-shaped molecules pack together to form a columnar liquid crystal phase with a characteristic defect texture.

under the umbrella of soft matter as they are weakly or unordered systems. They are, however, not equilibrium thermodynamic phases.

To think about self-assembly in this way, we can compare two different methods for depositing a molecular pattern on a surface. First, I could coat a substrate with positively charged ions in a particular pattern, then immerse the substrate in a solution containing a negatively charged polymer. The negatively charged polymer molecules move randomly in the solution until they encounter a positively charged area on the substrate, and stick. In this way, the polymer self-assembles into a patterned coating on the substrate by electrostatic attraction. Figure 2.2 shows an example of a polymer self-assembly formed in this way. An alternate strategy would be to directly

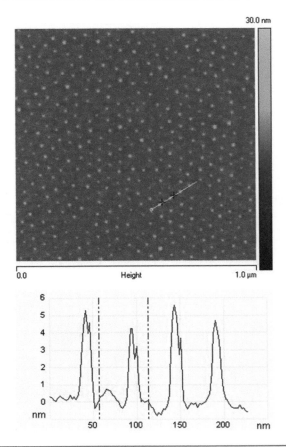

FIGURE 2.2   Contact mode atomic force microscopic (AFM) image of gold nanoparticles self-assembled on a silicon surface by block copolymer templating. Hexagonal patterning of the uniform-size nanoparticles can be seen in the image. Below the image, a height profile across several nanoparticles is shown, as indicated by the line marker on the image. (Courtesy of Chai Lor and Jennifer Lu.)

paint (by some clever method) the charged polymer onto the surface in the required pattern. This second method does not involve self-assembly as the coating is directed mechanically and not by thermal fluctuations and intermolecular forces. As you can imagine, self-assembly can be a very attractive way to obtain interesting two- and three-dimensional structures with little effort—if the molecules do what you want them to do!

In self-assembly processes, you will often hear mentioned that when a system reaches an equilibrium state, its *free energy* is minimized. By definition, the term free energy refers to the maximum amount of energy available

FREE ENERGIES AND ENTROPY

There are two common definitions for the free energy of a thermodynamic system, first, the Helmholtz free energy, F:

$$F = U - TS \qquad (2.1)$$

where U is the internal energy of the system, T is the absolute temperature, and S is the entropy. The Gibbs free energy, G is more commonly used in the chemical sciences and is given by the slightly modified formula:

$$G = U - TS + PV \qquad (2.2)$$

where P is the absolute pressure, and V is the volume of the system. As you can see, Gibbs free energy is quite similar to the Helmholtz free energy except for the additional PV term. This equation is more complete because it allows us to take into account the effects of volume changes on the system. Usually volume changes are significant where we have a phase change (say liquid to gas). In soft materials, most structural changes don't involve a significant change in system volume, but it is important to consider.

Entropy is a macroscopic thermodynamic variable that represents the number of possible distinct states (microstates) a system can have, for a particular measurable macrostate. The classic formulation is:

$$S = k_B \ln \Omega \qquad (2.3)$$

where Ω is the multiplicity of the system (see Appendix D for a more detailed explanation of entropy and multiplicity). Large thermal systems evolve until they reach their most likely state, thus maximizing their entropy. The term *entropy* can also be used casually (although not exactly correctly), to refer to the random thermal motions in a system, because increasing the temperature increases the entropy of a system. In general, its best to avoid using entropy as a measure of disorder.

From Equation (2.1), the Helmholtz free energy, we should picture the internal energy term U as being the energy it takes to create a system (e.g., the energy stored in the interatomic bonds and molecular motions). We subtract TS because this represents the energy spontaneously given to the system by its surrounding environment (in the form of heat, Q). This heat cannot be used to do work.

In real systems that undergo structural changes, we don't care too much about the absolute value of the free energy. Changes in free energy are more important. Using Equation (2.1) or (2.2), a change in the entropy of a system at constant volume, temperature, and number of particles is related to the change in the free energy by:

$$dF = -TdS \qquad (2.4)$$

From this result, we can see that any increase in entropy will correspond to a decrease in the free energy; when entropy is maximized at equilibrium, the free energy is minimized.

in the system that can be converted into work. A more useful way of thinking about free energy is to consider how it might change when we go from one material structure to another. For example, graphite and diamond are both materials formed from carbon atoms—but they have different free energies at standard temperature and pressure (STP). In fact, graphite has a lower free energy than diamond at STP but because these materials are both solids it's not easy for the carbon atoms to rearrange into their lowest free energy state! Given the ability to rearrange, systems spontaneously evolve to their lowest free energy state over time. For soft matter close to room temperature, usually the molecules or particles that form a phase are able to rearrange quite readily (for example, proteins in an aqueous solution or the soap molecules that form micelles around oily dirt). Soft materials can form a huge variety of different equilibrium phases—often with complex and surprising molecular structures—because their constituent molecules are able to rearrange and explore many different configurations.

2.2   INTERMOLECULAR FORCES

In Chapter 1, we looked at the behavior of molecules in an ideal gas. In such a system, the individual molecules that form the gas phase are so far apart on average that we assume they do not attract or repel each other. This description works well at temperatures significantly far from a phase boundary and for very dilute systems; however, in the case of other phases of matter (solids, liquids, or other condensed states found in soft matter), the forces between molecules become critically important and drive the structure of the phase. To discuss and understand the factors that determine the stability of different types of soft matter (complex fluids, amorphous solids, etc.), we must have a good understanding of the forces that can act between molecules. There are only a few fundamental forces in the universe, and in the world of soft materials, electromagnetism dominates. Essentially all of the intermolecular forces described here result from the electromagnetic force, but this interaction can manifest in a variety of interesting ways.

2.2.1   VAN DER WAALS ATTRACTION

Between any two atoms (and hence any particles placed close enough to each other), there is an attractive force known as the van der Waals force (or London force). This force results from a dipole–dipole interaction between any two atoms located close to each other. If we consider a simple view of two hydrogen atoms, each as a proton with an orbiting electron, one can see that each atom could be considered as a fluctuating electric dipole. The interactions between these dipoles when brought close together produce a net attractive potential that drops off as r^{-6} at large distances, where r is the dipole separation. The origin of this force can be derived using quantum mechanics, and although the derivation is beyond the scope of this book, you can find more detail in the excellent condensed matter physics book by Chaiken and Lubensky.[3]

2.2.2   HARD SPHERE REPULSION

As the two atoms are brought very close together, they experience a different, repulsive force known as the *hard sphere repulsion*. The origin of this force is simple to understand qualitatively. When two atoms become so close together that the volumes they occupy start to overlap, we expect a strong repulsion: Two atoms cannot be in the same place at the same time. This interaction has its origins in quantum mechanics; according to the *Pauli exclusion principle*, two electrons cannot occupy the same quantum mechanical state. Therefore, if two atoms are brought close enough together that their electron clouds begin to overlap, there will be a strong repulsion between the two atoms. This force is very short range.

The graph in Figure 2.3 represents a combination of the van der Waals attraction and the hard sphere repulsion. At very small atomic separations, there is a strong repulsive contribution to the curve. This dies away rapidly with separation, and the attractive force takes over. The *Lennard-Jones potential U_L* (Equation 2.5) is an empirical potential, often used to model short-range atomic interactions, and includes both repulsive and attractive terms,

$$U_L(r) = 4\varepsilon\left[\frac{r_0}{r^{12}} - \frac{r_0}{r^6}\right] \tag{2.5}$$

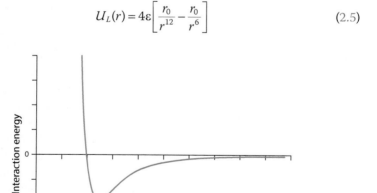

FIGURE 2.3   The Lennard-Jones potential plotted as the interaction energy between two particles as a function of separation distance r. This potential represents both the van der Waals attraction and the hard sphere repulsion.

In this formula, r is the interatomic separation, and ε is a constant that represents the potential at the energy minimum defined by $r = 2^{1/6}\, r_0$.

Because this potential includes positive and negative terms (to represent the attractive and repulsive forces between atoms), the curve has a minimum point, representing the minimum potential energy for the two-particle system. This point can be thought of as the equilibrium distance for the attractive and the repulsive forces.

2.2.3 ELECTROSTATIC FORCES

Very often, the basic elements of a soft phase will be charged, and the distribution of these electrical charges can lead to self-assembly. The attractive force between oppositely charged particles is described by the Coulomb force,

$$F_c = \frac{q_1 q_2}{4\pi\varepsilon_0 r^2} \tag{2.6}$$

where q_1 and q_2 are the charges of the two charged particles measured in Coulombs; ε_0 is a constant, the permittivity of free space; and r is the separation between the charges. The Coulomb potential is much longer range than the Lennard-Jones potential, and for charged solution-based systems will be a significant contributor to phase behavior. In the chapter on colloids, we will see examples of how electrostatics can play an important role in self-assembly. Charged particles or proteins in an ionic solution (where positive and negative ions are also present) can behave in interesting and non-intuitive ways.

2.2.4 HYDROPHOBIC EFFECTS AND THE HYDROGEN BOND

A large number of soft matter systems exist in aqueous solution, and therefore one of the most common driving forces for assembly is the *hydrophobic effect*. Water is a particularly interesting solvent, ubiquitous on Earth and vital to biological processes. Some of the properties of water are surprising. For example, the density of ice is lower than that of liquid water (hence ice floats), and the latent heat of vaporization is unusually high. These properties arise as a result of the polar nature of the H_2O molecules and the electrostatic bonds known as *hydrogen bonds* that they form with each other (see Figure 2.4). The hydrogen bond is considerably weaker than the covalent bond, but these short-lived bonds play a critical role in aqueous solutions. Hydrogen bonds actually give water a reconfigurable "structure" that has been the subject of considerable research. The formation of hydrogen bonds between water molecules is the origin of the *hydrophobic* and *hydrophilic* interactions that occur between different molecules and water.

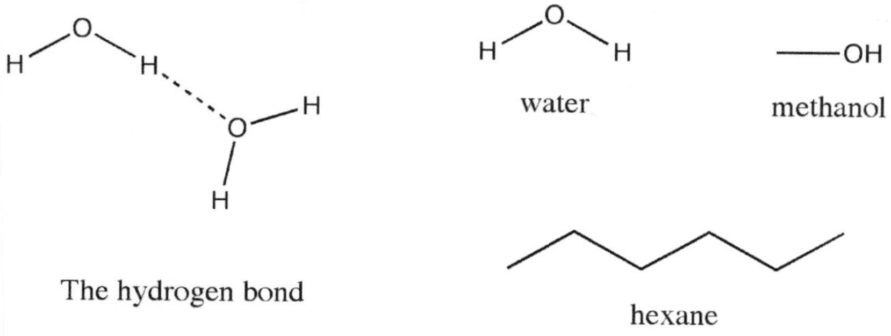

The hydrogen bond	water	methanol
	hexane	

FIGURE 2.4   The hydrogen bond and molecular structures for water, the polar solvent methanol, and the non-polar solvent hexane.

The hydrogen bond forms as a result of the polar nature of the water molecule. Close to the oxygen atom, there is a net negative charge, whereas the hydrogen regions have a net positive charge. This dipolar nature creates an electrostatic attraction between the hydrogen atoms and the oxygen atom in other molecules. Hydrogen bonds are relatively weak, with an energy of ~20 kJ/mol, meaning that they are much more easily broken than other kinds of atomic bonds.

If a different polar molecule is placed in water, it also may form hydrogen bonds with the water molecules and is therefore said to be hydrophilic. Non-polar molecules, however, will not form these bonds, so instead, the water molecules must arrange to form a cage-like structure around the non-polar species. The formation of this water cage is entropically unfavorable, as the addition of a non-polar molecule forces the water molecules to arrange in a more ordered state than they would if they were to enclose a polar molecule. This means that, there is a decrease in the local entropy of the system and a corresponding energy cost to incorporating the hydrophobic molecule into the solution.

As an example, let's consider oil and water. The hydrocarbon chains of an oil molecule are non-polar, so when these chains are placed into water, they surround the hydrocarbon chain (but don't hydrogen bond with it). This effect limits how many ways water molecules can arrange to surround the carbon chains. Mixing oil and water on the molecular level would produce a decrease in the net entropy of the system (with a corresponding increase in free energy), and so on the macroscale, we observe that oil and water do not readily mix—their equilibrium state is to phase separate. A better solvent for oil would be the non-polar hexane, whereas the polar methanol would not be effective. Molecular structures for the molecules mentioned are shown in Figure 2.4

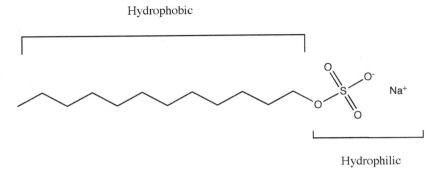

FIGURE 2.5   An example of an amphiphilic molecule, sodium lauryl sulfate (SDS). This common detergent, found in most shampoos, has a hydrophilic head group and a hydrophobic tail.

Certain molecules exhibit both hydrophobic and hydrophilic properties; these are known as amphiphilic. An example of a familiar amphiphilic material is ordinary soap. Soap molecules (or detergents in general) have a hydrophilic and a hydrophobic part as shown in Figure 2.5 for a common household detergent. When these detergent molecules are dissolved in water, they spontaneously come to an arrangement such that the hydrophobic hydrocarbon chains are located as far away from the water as possible. This tendency leads to a rich phase behavior and an effective mechanism for separating out oily dirt. We will be discussing the action of detergents in some more detail in Chapter 4. The hydrophobic effect is very important in nature. Waxes are very hydrophobic and provide a mechanism for many plants to repel water droplets from their surfaces (Figure 2.6). Hydrophobicity is also important on a cellular level, in particular for the cell membrane and membrane proteins. Membrane proteins are large molecules that locate across the cell membrane—including the hydrophobic core. They need to have a particular hydrophobic/hydrophilic surface to do so.

2.3   AGGREGATION AND ASSEMBLY

This book is all about materials that are soft in nature, but the way in which the constituent molecules in many soft materials are arranged is not random. Instead, they have some structure that can be measured, even if they do not exhibit long-range order like a crystalline solid. The molecules or particles that make up a material are subject to attractive and repulsive forces from each other and any solvent molecules in the system. As a result of these interactions, spontaneous assembly may occur. The processes that lead to this self-assembly can be thought of in two different ways:

FIGURE 2.6 The waxy coating on the surface cuticle of many leaves and flowers produces a hydrophobic surface. Water beads on the surface and runs off easily.

1. Formation of a thermodynamically stable uniform phase or
2. Aggregation of a structure by some diffusive process.

We have already discussed the first of these by considering the solid, liquid, and gas phases. These phases, among others, spontaneously form as a result of the competing effects of thermal fluctuations and intermolecular forces, leading to a uniform material structure. A variety of interesting thermodynamically stable, reversible phases will be introduced throughout this book as we describe the diverse forms of soft matter. In particular, there are many stable phases that do not fall neatly into the classifications of solid, liquid, or gas.

The second mechanism for self-assembly depends on exactly how the constituent particles come together (i.e., the method of assembly), and the structures that form as a result may be reversible or irreversible. Many materials can be thought of as aggregates, rather than in a thermodynamic phase, and these materials are usually assembled by a growth process. In a simple liquid, the constituent molecules are free to diffuse throughout the system. In theory, every molecule can explore every possible position in the system. Given enough time, the material will adopt its most stable (equilibrium) state. In contrast, many soft systems are formed by an alternate assembly mechanism, in which structure develops by adding material to a growing interface. For example, small particles suspended in a fluid tend to stick to each other. A particle cluster will gradually grow as more and more particles add to its surface over time producing an internal structure very different to the structure which would form in an equilibrium assembly process. As these clusters grow, voids of different sizes randomly appear in the structure. It becomes unlikely that any enclosed holes will be filled as newly added particles tend to stick on the surface. This assembly produces an interesting soft material with a non-regular, loose internal structure. In Chapter 6 we will look at this more closely for colloidal assemblies.

2.3.1   POWER LAWS AND FRACTALS

The way that materials fill space can be more complicated than you may first think. We are used to treating objects as uniform in density, but the concept of uniformity in real materials often depends on how closely you look. A uniform three-dimensional object, such as a perfect solid crystal is translationally invariant, free from holes or gaps in the crystal structure. This means that if you sit on a point in the crystal lattice and look around, your surroundings will appear the same from every point. Many materials, however, do not have this idealized structure and may be full of holes of all different sizes and shapes. In this case, where you sit in the material will affect the view around you. This volume "roughness" is an important feature of most real objects because not many materials are perfectly homogeneous. The idea can also be applied to just two dimensions (2D) because the outline of different solid objects may also have a complex surface texture. Few materials are atomically smooth on their surfaces; indeed, on close inspection most are rough and pitted.

If a straight line is one-dimensional and a flat sheet is 2D, how can we describe the dimensionality of a coastline or the shape of a polymer

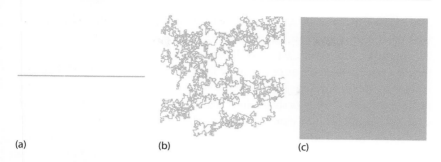

(a) (b) (c)

FIGURE 2.7   Examples of varying dimensionality: (a) a straight polymer; (b) a self-avoiding polymer (2D); and (c) a flat plane.

molecule? In these cases, 2D or 3D space is being partially filled by a 1D object. Figure 2.7 illustrates this concept; the coiled up self-avoiding polymer wraps around on itself to partially fill the 2D space. The same concept can be applied to the boundary between two and three dimensions. Take a flat sheet of paper, then crumple it up in your hands; the 2D object occupies a three-dimensional space, but incompletely.

We can describe these complex geometrical shapes mathematically, and if the space-filling properties of the objects are consistent over a wide range of length scales, the system can be characterized using a quantity known as the *fractal dimension*. The concept behind the fractal dimension is that these objects are somewhere between the integer dimensions—they have a "fractional dimension." By calculating the fractal dimension we can obtain a measure of how space is filled by the material.

Let us think about this concept in terms of filling space with mass by returning to the examples in Figure 2.7. Take the straight polymer in the Figure 2.7a, then pick an origin at the center of the molecule, and calculate how the total mass enclosed by a circle of radius r increases as a function of r. You should obtain a linear function,

$$M(r) = Ar \tag{2.7}$$

where A is a constant.

If we now consider the solid sheet, completely filled in two dimensions, and repeat the same process, we can see that the mass scales as:

$$M(r) = Ar^2 \tag{2.8}$$

Finally, if we calculate how mass scales with radius for our self-avoiding 2D polymer example, we say very generally that:

$$M(r) = Ar^D \tag{2.9}$$

The exponent D is defined as the fractal dimension and should lie somewhere between 1 and 2 in this example.

In this simple case, we have looked at a polymer over just a limited length scale; however, one of the most fascinating aspects of fractal objects is their tendency toward *scale invariance*. In other words, if you were to look at an object, observing its shape and structure, then zoom in on a particular spot, you would observe exactly the same shapes and structure. This sounds confusing, but actually we observe objects with these properties every day. Look closely at plants and trees, and you will observe a characteristic branching geometry. Tree branches stick out from the trunk at a particular angle, then each of these branches has smaller twigs protruding at the same angle, and so on. You could also think about a classic cloud shape. Can you tell how large a cloud is with no frame of reference? These examples demonstrate scale invariance, and there are also many examples of this kind of structure in soft materials, particularly in colloidal and polymer aggregates.

Now, imagine a tree-like branching motif that fills space by constantly branching from smaller and smaller limbs; as we go to higher and higher generations of branching, the space between the largest limbs will be gradually filled but voids will also persist. Simple fractals can be easily generated by this kind of repeating algorithm. Two well-known examples are shown in Figure 2.8: the box fractal and the Sierpinski gasket. Even if the algorithm is repeated forever, the 2D space occupied by the structure will never be completely filled.

The fractal dimension gives a measure of the extent to which space is filled and is often used to help characterize materials, for example, the arrangement of polymer chains in solution. You may come across this

(a)

(b)

FIGURE 2.8   Two well-known examples of methods to generate simple fractals: (a) the box fractal and (b) the Sierpinski gasket. In these examples, a simple algorithm is repeated, creating finer and finer structures.

parameter expressed as the exponent of a *power law* behavior, where the equation of mass distribution $A(r)$ scales as:

$$A(r) = Br^D \text{ so } \log A(r) = \log B + D \log r \tag{2.10}$$

where B is a constant.

For the most part, calculating a fractal dimension for a material is only interesting if the value obtained is constant over a range of length scales. In other words, the pattern of organization must repeat itself sufficiently for the material to be structurally scale invariant. Still, calculation of D across different ranges provides a powerful technique for characterizing the internal structure of a material.

One simple way of determining the fractal dimension from an image of a material is to use a "box-counting method." The fractal is covered by a grid of boxes of size x and the number of boxes containing mass (N) counted. Finer and finer grids are used, and N is plotted as a function of x, resulting in a power law behavior that can yield the fractal dimension. This method is demonstrated in Figure 2.9 for a coastline.

The fractal dimension for a material can also be determined experimentally using scattering (X-ray, neutron, or light scattering are common), and this method is discussed briefly in Chapter 5.

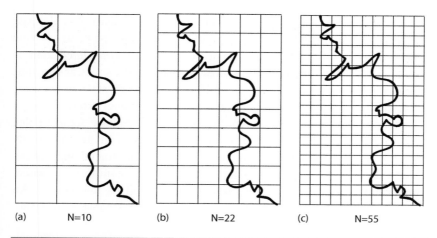

(a) N=10   (b) N=22   (c) N=55

FIGURE 2.9 Coastlines are examples commonly used for demonstrating scale invariance. If you try to measure the length of a coastline with a measuring stick, the shorter you make the stick, the greater your answer will be. This fact is also captured in the box-counting method for determining fractal dimension. Boxes of successively decreasing size r are placed over the coastline, as shown in (a–c). The number of boxes crossed by the coastline N is recorded in each case. The slope of a log/log plot of N(r) will give the scaling dimension.

2.3.2 AMORPHOUS MATERIALS AND SHORT-RANGE ORDER

In hard crystalline materials, the constituent atoms are very well ordered into a lattice. We describe crystalline solids as having long-range order. If I sit on an atom in a crystal lattice, I can confidently predict the positions of the surrounding atoms in the lattice over long distances by knowing the lattice parameters (the characteristic dimensions of the unit cell). Soft materials are typically quite different. They are strongly influenced by thermal fluctuations and therefore exhibit only short-range order at best. Think about the distribution of molecules in a fluid. Starting at a random molecule in the fluid, you might be able to make a reasonable guess for the distance to a neighboring molecule, but trying to predict the position of the next-nearest neighbor is very difficult. Fluids have no molecular positional order and intermolecular distances are not very well defined. Some soft materials are quite rigid, but still have little to no positional order—they are amorphous. For example, the amber glass in Figure 1.3 and hard plastics (polymer glasses). There are also classes of materials which exhibit some short-range positional and/or orientational order, and these liquid crystals will be discussed in detail in Chapter 3.

2.4   ACTIVE MATTER: BEYOND EQUILIBRIUM ASSEMBLY

One of the most exciting developments in soft matter science over the past few years has been the rapid expansion of research into the relatively new field of *active matter*. Up to this point, we have only considered materials which are either in equilibrium already (e.g., liquids, solids at constant temperature), or that are at least assembled into a steady state structure—a material in equilibrium will tend to stay that way unless external energy is somehow injected into the system (heating to melt a phase, for example). In the previous section, we also introduced materials which are in a steady state, but not in their lowest energy configuration (e.g., fractal aggregates). We can think of these systems as being trapped out of equilibrium—where equilibrium represents a system that has minimized its free energy. An aggregate of particles is not exactly "solid," but remains constant in structure once formed.

A third category of material—active matter—is somewhat different. Active materials are materials which do not maintain a steady state. The unifying theme of active matter is that their basic subunits (i.e., molecules or particles) take in energy locally from the environment, and then translate that energy into movement. The local movement of the particles can produce large scale structural dynamics within the material.

For example, imagine a system of particles in a box, (similar to our earlier ideal gas model). Instead of passive particles, moving randomly, the particles are self-propelled (or active). Adding in an "active" behavior to the particles can totally change the structure of the material. For example, active colloids (self-propelled micron scale particles) have been observed to cluster together and "swarm" or form different self-assembled arrangements as they move collectively. This behavior is completely different from the ideal gas! and other equilibrium phases.

There have been many examples of this class of material observed in biology and active behavior can be observed across a wide range of length-scales, from the molecular to the macroscopic. In this book, we will learn about a few examples of these fascinating materials and look at examples in biological systems, ranging from bird flocks and insect swarms, to cells and biopolymers with molecular motors (Chapter 7).

2.5 MECHANICAL PROPERTIES OF SOFT MATTER

When we introduced the concept of soft matter at the beginning of this chapter, different types of household materials were given as examples. It is impossible to classify common materials like Jell-O, toothpaste, or soap as a simples solid or liquid since they fall somewhere between the two. These and other soft materials are easily deformable—their mechanical properties are different from a solid, but neither are they exactly liquids. If we want to describe different types of soft matter in terms of their mechanical properties, then we need to review our definition of softness. The ability of a material to deform when subjected to an applied force can be described in terms of measurable physical parameters such as *stress* and *strain*.

2.5.1 VISCOSITY AND ELASTICITY

The stress on a material, σ, is defined as the applied force per unit area,

$$\sigma = \frac{F}{A} \tag{2.11}$$

Tensile stress results when we pull on an object, trying to stretch it, and compressive stress results from a force in the opposite direction, when we try to compress the material. In each case, the force is applied perpendicular to the cross-sectional area A as shown in Figure 2.10.

An *elastic* material will respond to a tensile stress by stretching in a linear fashion according to Hooke's law,

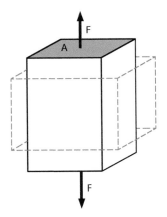

FIGURE 2.10   A material under tensile stress with a positive Poisson's ratio.
The dashed gray lines indicate the original shape of the material.

$$F = -k\Delta x \qquad (2.12)$$

where F is the restoring force, Δx is the change in length in the direction of
the force application, and k is the elastic modulus. *Shear stress* (Figure 2.11)
results when a force is applied to a material parallel to the cross-sectional
area A.

We define the strain ε on a material as:

$$\varepsilon = \frac{\Delta l}{l} \qquad (2.13)$$

where l is the original length of the material, and Δl is the change in length
after the application of a force.

Stress and strain are related by the Young's modulus Y, where:

$$Y = \frac{\sigma}{\varepsilon} = \frac{Fl}{A\Delta l} \qquad (2.14)$$

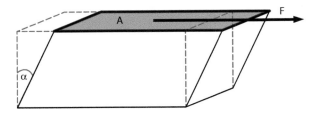

FIGURE 2.11   A material under shear stress. Dashed lines indicate the original
shape of the material.

By rearranging this formula to resemble Hooke's law, we can see that:

$$F = \frac{YA\Delta l}{l} \tag{2.15}$$

so the elastic modulus defined in Hooke's law k is related to the Young's modulus by:

$$k = \frac{YA}{l} \tag{2.16}$$

Elastic materials deform linearly according to Hooke's law, and as they are stretched their constituent molecules become displaced from their equilibrium positions. A simple mechanical model can represent the behavior of a purely elastic material as a lattice of balls connected by springs (Figure 2.12). This elastic behavior holds until the applied force is large enough that the material begins to deform irreversibly. Ductile materials exhibit a plastic deformation regime at stresses above the yield stress, whereas a brittle material will fracture at that point.

A typical elastic material will respond to a tensile stress by elongating in the direction of the force applied, and contracting laterally. We describe this behavior as having a positive *Poisson's ratio v*. Poisson's ratio is the ratio of the transverse strain perpendicular to the direction of the force and the linear strain in the direction of the force.

$$\upsilon = \frac{-\varepsilon_\perp}{\varepsilon_\Pi} \tag{2.17}$$

Most solids have a Poisson's ratio of approximately 0.3, although if we include soft materials, this range is much larger (Figure 2.13). Some typical Poisson's ratios for different materials are listed in Table 2.1. Interestingly, there are

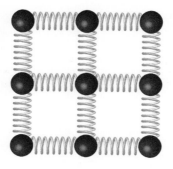

FIGURE 2.12   The atoms of a solid under small deformations can be modeled as balls connected by simple springs that obey Hooke's law.

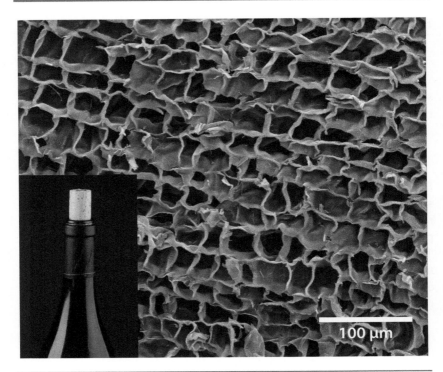

FIGURE 2.13   Cork is an interesting material[2]: Its cellular structure results in a Poisson's ratio of nearly zero. This means that when the cork of a wine bottle is compressed, it does not expand laterally, making it easier to use as a stopper that can be reinserted into the bottle by pushing down. The image on the right shows a scanning electron microscopic (SEM) image of the interior structure of a wine cork. (Courtesy of Chi-Shuo Chen and Prof. Wei-Chun Chin.)

TABLE 2.1
Young's Modulus and Poisson's Ratio for a Selection of Different Hard and Soft Materials

Material	Young's Modulus (N/m²)	Poisson's Ratio
Borosilicate glass	63×10^9	0.2
Copper	11×10^{10}	0.37
Lead	16×10^9	0.45
Rubber	5×10^7	~0.5
Polystyrene	3.1×10^9	0.1–0.4
Cork	15×10^6	~0

FIGURE 2.14   Honey is a good example of a purely viscous material.

certain materials that have a negative Poisson's ratio.[1] In such a case, a material under tensile stress will actually become wider in the transverse direction.

In the case of a liquid, there are no springs connecting the balls: Liquids do not exhibit elastic behavior; they are *viscous*. In the viscous regime, adjacent molecules are able to slip past each other easily when a shear stress is applied. Viscous materials can be poured and spread easily, they flow to adopt the shape of their container and do not retain a specific shape (Figure 2.14).

Imagine a simple scenario. Two parallel plates enclose a layer of liquid. The top plate is moved to the right long the x-axis at a constant velocity, applying a sheer stress to the liquid. We define a *Newtonian fluid* as one for which the strain rate is constant as we apply this constant sheer stress. This relationship is quantified by the following equation:

$$F = \eta A \frac{du}{dy} \tag{2.18}$$

F is the applied force, and *A* is the area as pictured in Figure 2.11. Here, *u* is the velocity of the sheared plate, so *du/dy* is the velocity gradient in the fluid perpendicular to the sheared plate, or the *strain rate*, and n is the viscosity of the liquid.

Soft materials are often *viscoelastic*: They can exhibit both elastic and viscous behavior, depending on the intermolecular potentials present in the system. In Chapter 5, we look at how the viscoelastic properties of soft materials are measured in the case of polymers. Many soft materials have a complex, anisotropic structure. This means that the elastic properties depend on the direction of applied stress. Fabrics provide a good example of this phenomenon. For example, stretch denim is more elastic horizontally than vertically.

The viscous or elastic response of a material may also vary depending on the timescale over which a shear is applied. Some materials exhibit elastic properties when sheared rapidly, but then become viscous when poured or sheared slowly. An interesting example of this effect can be observed using a thick mixture of cornstarch (corn flour) with water. This simple mixture exhibits different viscoelastic behaviors depending on shear rate. If the mixture is prepared to the correct concentration, the resulting goo will flow like a liquid when poured slowly, but behave like a solid and even break when deformed rapidly. This behavior stems from the microstructure of the starch/water mixture; starch particles cluster and jam at high shear rates, creating a solid-like response.

QUESTIONS

1. What is the difference between a ductile and a brittle material? How will their stress/strain curves differ?

2. A typical Young's modulus for a perfect crystal lattice of gold should theoretically be about 10^{10} Nm^{-2}; however, experimentally values between 10^6 and 10^9 Nm^{-2} are observed. Why do you think this is?

3. A solid plastic brick with dimensions $a \times b \times c$ is compressed by a force applied parallel to a, and a volume change is measured. Show that the change in volume under an applied force parallel to a is equal to:

$$\Delta V = (1 - 2v)\Delta abc$$

where v is Poisson's ratio, and ΔV is the change in volume. What is the significance of a Poisson's ratio of 0.5? Give an example of a common material with this property.

4. Rubber bands are made from a stretchy polymer material. If a section of a rubber band with an unstretched cross-sectional area of 0.4 mm² increases in length by 50% under an extension force of 15 N, calculate its Young's modulus. How does the value you obtain compare to typical Young's moduli for solid metals or ionic crystals (e.g., sodium chloride).

REFERENCES

1. R. Lakes, Foam structures with a negative Poisson's ratio. *Science* 235, 1038–1040 (1987).
2. L.J. Gibson, K.E. Easterling, and M.F. Ashby, The structure and mechanics of cork. *Proc. R. Soc. Lond. A* 377, 99–117 (1981).
3. P. Chaiken and T. Lubensky, *Principles of Condensed Matter Physics*, 2nd ed. Cambridge, UK: Cambridge University Press (2006)

Liquid Crystals

3.1   INTRODUCTION TO LIQUID CRYSTALS

3.1.1   WHAT IS A LIQUID CRYSTAL?

When introducing the subject of liquid crystals, the first and most important question to be answered is, what *is* a liquid crystal? Unlike the more well-known varieties of soft matter, such as polymers or gels, liquid crystals are often something of a mystery to the non-specialist, although a simple definition can be made for the wide variety of materials that can exhibit liquid crystalline properties.

A liquid crystalline phase is an ordered fluid phase with some degree of *anisotropy*. This state of matter occurs in certain materials between the more familiar solid and liquid phases. It is best to think of a material as *exhibiting* a liquid crystalline phase rather that material "being a liquid crystal," as all liquid crystal materials can be cooled and "frozen" to form a solid phase or heated and "melted" to form a liquid phase. There are many different liquid crystal phases that can occur between the traditional solid and liquid phases. A given material may exhibit several liquid crystal phases with quite different structural, electrical, and optical properties.

Traditionally, the discovery of liquid crystals has been attributed to Friedrich Reinitzer in 1888. Reinitzer was a scientist at the German University of Prague in the Institute of Plant Physiology; while studying the properties of cholesterol compounds, he found that cholesterol benzoate appeared to have two different melting points.[1] Another scientist of the time, Otto Lehmann, a crystallographer at the University of Aachen, first proposed the liquid crystal state to be a distinct new state of matter.[2]

Liquid crystals are now best known for their applications in liquid crystal displays (LCDs), but in fact the kind of liquid crystal used in such displays represents a small sub group of the family of liquid crystal phases and materials. In this chapter, we will learn about the main classifications of liquid crystals, their properties, and their applications, including a description of standard experimental techniques for phase identification.

3.2   ANISOTROPY IN LIQUID CRYSTALS

The term *anisotropic* is often used in physics to describe a system with some form of directionality to its physical properties. The liquid crystalline state is described as being anisotropic. Typically, this property results from the packing of non-spherical molecules or molecular assemblies (e.g., rod-like, disc-like, etc.). It may also result from the packing of molecules that are spherical in shape but have some other form of anisotropy; that is, they are more charged on one face or more hydrophobic/hydrophilic in some region.

If the molecules (or molecular assemblies) that make up the material exhibit some form of anisotropy on a molecular level, then in a bulk phase their equilibrium packing configuration at a given temperature may also be anisotropic, resulting in a liquid crystalline phase. It is important to note here that the liquid crystalline phase is a macroscopic state of matter, resulting from molecular-scale interactions. On heating to a high enough temperature, this packing may no longer be favored, and on melting to the liquid phase, even anisotropic molecules will give an isotropic bulk phase. Figure 3.1 demonstrates the idea of how bulk anisotropy can be manifested by anisotropic molecules. In the first image on the left of Figure 3.1, we see spherical (isotropic) molecules forming a fluid-like phase. This phase will have the same properties when viewed or experimentally probed from any direction; therefore, the isotropic particles have resulted in an isotropic phase. The middle image shows an isotropic phase formed from anisotropic (in this case rod-like) molecules. Even though on a molecular level the material is anisotropic, bulk properties will be isotropic as this material will also appear the same from every direction. Any directionality from the individual molecules is averaged over all directions, and the molecules do not show any specific alignment. The first two phases here could represent the standard liquid phase (also known as the isotropic phase). The third image (on the right) demonstrates how rod-like molecules can give rise to an anisotropic bulk phase. By packing with a specific orientation, the physical properties of the phase will differ depending on the direction from which they are addressed. For example, imagine viewing this third phase under the microscope. It will appear differently when viewed from the side or from above. In fact, when viewed from above in this diagram, the phase will still appear to be isotropic, although a side view reveals the anisotropy.

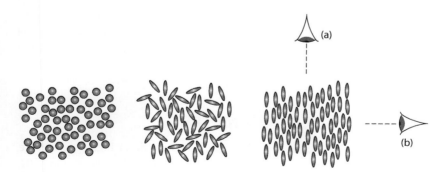

FIGURE 3.1 Demonstration of bulk anisotropy and isotropy. The left and center images show molecules in an isotropic phase, despite differences in molecular shape. The phase on the right can appear isotropic when viewed from (a) but anisotropic when viewed from (b).

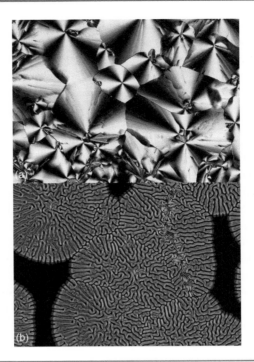

FIGURE 3.2 Examples of liquid crystal birefringence textures imaged at 100X magnification: (a) a homeotropic texture of a columnar phase and (b) "fingerprint" texture of the cholesteric phase growing in from the isotropic liquid phase (black regions).

Anisotropy and its relationship to molecular orientation is extremely important in the study of liquid crystal materials. Careful examination of a microscope image combined with some knowledge of the structure of the liquid crystal phases can allow us to deduce local molecular orientation (or alignment direction), characterize phase behavior, and gain a better understanding of the optical and electromagnetic properties of a material (Figure 3.2). These techniques are discussed in the following section along with a variety of different microscope images for common liquid crystal phases.

3.3 THE ORDER PARAMETER

The orientation of a liquid crystalline phase can be described by a vector **c**, the director. This vector is used to indicate the average local orientation of the molecular long axis **a**. The director describes the average molecular orientation on a length scale much larger than an individual molecule, but typically less than the length over which deformations of the phase can occur. The vector **c** describes the average orientation of molecules in the phase but gives no information on the degree of ordering; for this we use S, the order parameter. For a perfectly aligned phase in which all molecules point in the same direction, $S = 1$, and for a phase with no preferred direction at all, $S = 0$.

Unlike a solid crystal, in which all molecules in the lattice are consistently oriented to form the crystal structure, in the liquid crystal phase there is a significant amount of freedom per molecule. The phases are fluid-like in that molecules are not confined to lattice positions and may diffuse throughout the bulk. They are, however, subject to certain packing constraints.

When considering the bulk properties of a liquid crystal material, the liquid crystal molecule is usually assumed to be symmetric about the **a** axis, and molecular orientations in the **a** and −**a** directions are also considered equivalent (although details of different molecules will of course vary). The molecules are often depicted as simple shapes (rods, disks, etc.), ignoring all molecular details. In this book, we will follow this convention, with the addition of flexible chains as this represents the most common structure for thermotropic liquid crystals: a rigid, rod-like core with flexible alkyl chains on either end. It is the molecular details that result in the specific phase sequence for a given material, so a great deal of emphasis is given to details in the design and synthesis of new materials. However, for general phase descriptions, a simplified model can be used.

3.4 THERMOTROPIC AND LYOTROPIC LIQUID CRYSTALS

Liquid crystal materials can be grouped into two main classifications, thermotropics and lyotropics. A thermotropic phase is one that can form by heating or cooling a material. Just as we see a phase transition between solid and liquid as we heat and melt ice, in thermotropic liquid crystals, additional "melting points" can be observed in between the solid and the liquid phases. These are the thermotropic liquid crystalline phases. A lyotropic liquid crystal phase is formed by molecules dissolved in a solvent, and phases form at certain concentrations in that solvent. In this chapter, we focus on descriptions of the thermotropic liquid crystal phases. Lyotropics, although also liquid crystals, are described in detail in Chapter 4, focusing on surfactants.

3.5 BIREFRINGENCE IN LIQUID CRYSTALS

Birefringence is an optical property of certain materials in which light rays of different polarizations are subject to different refractive indices when passing through the material. Many crystalline solids have this property; some well-known examples are calcite, tourmaline, and quartz. In birefringent crystals, the refractive index of the material is dependent on the polarization

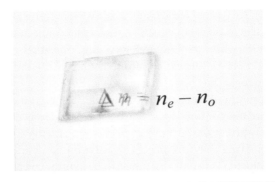

of the incident light with respect to the axis of anisotropy. Light with a linear polarization perpendicular to the axis of anisotropy is subject to the ordinary refractive index n_o, and light with a linear polarization parallel to the axis of anisotropy is subject to the extraordinary refractive index n_e. We can see the birefringence effect demonstrated in Figure 3.3. A double refraction image appears because in this crystal the two different polarizations of light travel through the crystal at different speeds.

In liquid crystal materials, birefringence occurs similarly and originates from bulk anisotropy of the fluid. When a non-polarized monochromatic light ray passes into a liquid crystal medium, it can be resolved into two components with orthogonal polarization directions. If each of these rays experiences a different refractive index, and one travels faster than the other, creating a phase difference between the two.

The birefringence of a material, Δn, is given by,

$$\Delta n = n_e - n_o \tag{3.1}$$

and the phase difference by:

$$\Delta\phi = \frac{2\pi}{\lambda} d\Delta n \tag{3.2}$$

where d is the thickness of the liquid crystal slab, and λ is the incident wavelength. $d\Delta n$ is the optical path difference between the two orthogonally polarized waves, so the thickness of the slab is an important parameter; the further the two resolved light rays travel through the birefringent medium, the greater the phase shift between them until the phase shift is equal to 2π again. Birefringent materials can be used to generate any polarization state by selecting the correct sample thickness and incident polarization.

WORKED EXAMPLE 3.1

A liquid crystal film can be used as a quarter-wave plate to produce circularly polarized light. To achieve this, the liquid crystal material needs to provide a phase shift of a quarter wavelength ($\lambda/4$) between the ordinary and extraordinary waves when they exit the film.

If linearly polarized light is incident on a LC film, the incident electric field can be separated into two components, perpendicular and parallel to the **c** director.

$$E_1 = E_0 \; sin(kx - \omega t) \text{ representing the ordinary wave}$$

$$E_2 = E_0 \; sin(k(x + \Delta x) - \omega t) \text{ representing the extraordinary wave}$$

On exiting the medium, the phase shift between these two waves, $\Delta\phi = k\Delta x$.

The difference in optical path length, between the two waves, $\Delta x = d(n_e - n_o)$, produces a phase shift of $\Delta\phi = k\Delta x$ and since $k = \frac{2\pi}{\lambda}$

$$\Delta\phi = 2\pi d(n_e - n_o)/\lambda$$

To produce a quarter-wave plate, $\Delta\phi = \pi/2$. Using a birefringent material with $\Delta n = 0.02$ at a wavelength of 400 nm, our film thickness must be equal to,

$$d = 400/4(0.02) = 5{,}000 \text{ nm} = 5 \text{ μm}$$

As this particular liquid crystal film is a quarter-wave plate, incident linearly polarized light will be converted into circularly polarized light on exiting the 5 μm thick film.

So far in this discussion, we have only considered monochromatic incident light. Illumination with white light, however, significantly complicates the situation. As we saw, the phase difference developed as an unpolarized light ray passes through a birefringent medium is dependent on wavelength. Because different wavelengths emerge from the liquid crystal slab with different polarization states, different colors (birefringence colors) can be observed if we place a second polarizer (analyzer) orthogonal to the first and look at transmitted light through the whole system. For light exiting the liquid crystal, the magnitude of the polarization component aligned with the analyzer will vary as a function of wavelength. Wavelengths with a large component aligned with the analyzer will be more intense than others that allow only a smaller component to pass through the analyzer. This effect creates the observed colors that depend on slab thickness and birefringence. One of the most powerful and simple techniques in the characterization of liquid crystal phases is polarized optical microscopy (POM). Using this technique, we can take advantage of the birefringence of liquid crystals to identify different phases. The sample is placed between crossed polarizers on the optical microscope and viewed in transmission. The illuminating light passes through a linear polarizer, the liquid crystal film, and then a second linear polarizer crossed with the first. This technique is described in detail in Section 3.8.2.

3.6   TOPOLOGICAL DEFECTS AND DEFECT TEXTURES

To have a discussion about the characteristics of various liquid crystal phases and how they appear under the microscope, we must introduce the topics of *topological defects* and *defect textures*. In the previous section, we saw that liquid crystal phases can be birefringent. They are fluids composed of anisotropic molecules (rods, discs, etc.) which tend to locally line up with each other. This behavior means that when the liquid crystal is oriented correctly between crossed polarizers, we expect to see transmission of light through the polarizers (more details on this later in the chapter). Liquid crystal molecules are typically quite small (~2–20 nm in length); therefore, using optical microscopy alone, it is not possible to directly observe the molecular packing structure in the phase (the resolution limit of an optical microscope is only about 200 nm.) The fluid-like nature of the phases also makes techniques such as atomic force microscopy (AFM) and electron microscopy (EM) unsuitable for studying molecular packing, except in the case of freeze fracture microscopy.

A perfectly oriented liquid crystal film in which all the constituent molecules are, on average, pointed in the same direction (aligned) over macroscopic length scales (~100 μm or more), should appear as an image of uniform brightness under a polarized microscope. The brightness and color of the transmitted light provides information on liquid crystal film thickness and director orientation; however, in this state, it is not necessarily possible to distinguish between phases easily, especially when trying to identify an unknown phase. Whenever a phase is not perfectly aligned, however, we typically see *defects*. These are places where the ideal molecular ordering of the phase is disrupted within the fluid. You might be quite surprised to learn that a fluid can have defects analogous to those seen in a crystal lattice, but liquid crystals are *ordered fluids* and retain some characteristics of the solid state as a result of their molecular anisotropy. Defects lend a characteristic appearance to a phase, aiding identification. They can be annoying (if you want a perfectly aligned material) but are often very useful in optical applications and liquid crystal composite materials where defects can be used to trap included particles or molecules.

3.6.1 INTRODUCTION TO TOPOLOGICAL DEFECTS

It is not typical for the molecules of a phase to align in one consistent direction over large distances unless they are forced to do so by applying a specific coating (*alignment layer*) to the confining substrates (e.g., glass plates). Instead, the molecular orientation will typically vary spatially over length scales much larger than the molecular scale, resulting in a variable birefringence across the material. There may also be point or line defects present in the packing structure (also known as *disclinations*). These defects and the resulting local variations in birefringence can be used to identify different liquid crystalline phases using the polarizing optical microscope.

Topological defects in liquid crystal are areas where the liquid crystal director cannot be clearly defined. Figure 3.4 illustrates some examples of

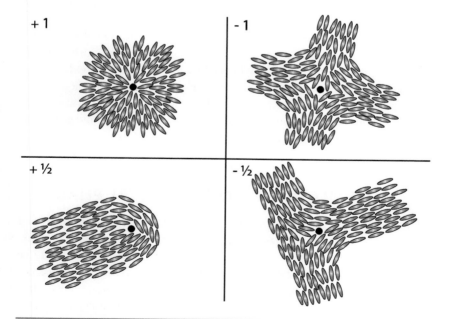

FIGURE 3.4 Four examples of different topological defects in a liquid crystal phase, including their topological charge, +1, −1, +1/2 and −1/2.

possible two-dimensional defects in a liquid crystal phase formed from rod-like molecules. At the core of each defect, we see there is a point where the liquid crystalline ordering breaks down, i.e., there is no specific molecular direction. Such defects can be induced by impurities or particles in the liquid crystal, but they can also occur spontaneously as the phase forms.

The numbers next to each defect in Figure 3.4 represent their topological charge—a method to classify different defects. To obtain the charge of a defect, simply follow a circular path 360° around the core and count how many rotations the director makes as you travel around—try it for each defect.

The microscopic images in which characteristic defects and birefringence patterns are seen are known as *defect textures.* Figure 3.2 shows two examples of the rich variety of defect textures that can be observed for the many different liquid crystal phases so far observed. The features we see in these polarized microscopy images result from defects in molecular packing and the changing orientation of the liquid crystal director. When placed between glass plates (or another substrate), liquid crystal molecules can be encouraged to orient themselves with a planar (homogeneous) alignment (the molecular long axis lies parallel to the substrate) or with a homeotropic (vertical) alignment (the molecular long axis lies perpendicular to the substrate). To achieve control of molecular alignment, specific coatings can be applied to the glass. A homeotropic alignment is usually optically anisotropic and therefore will not produce any birefringence (recall Figure 3.1a), but a planar alignment (Figure 3.1b), will produce a striking defect texture.

In the following sections, we will examine several common liquid crystalline phases and look at their defect textures when confined between glass and viewed on the optical polarizing microscope, discussing how particular characteristic defects are formed and how we can use them to identify the phases.

3.7 THERMOTROPIC LIQUID CRYSTAL PHASES

A large number of thermotropic liquid crystal phases have been identified and classified. Their molecular structures are varied and range in complexity. However, they all exhibit common properties of anisotropy and fluid-like behavior in at least one dimension. In the following sections, we discuss the most commonly encountered liquid crystal phases, describing their molecular structures and key properties. Novel, interesting liquid crystal phases are still being discovered as new materials are synthesized and investigated. The following sections focus on the most commonly encountered classifications: the nematic and smectic phases.

3.7.1   THE NEMATIC PHASE

The nematic phase is the simplest of the thermotropic liquid crystalline phases and is characterized by just one important feature: all of the molecules in the material tend to align in the same direction (indicated by the $\bar{c}$ director; Figure 3.5). You might be curious to ask, why do they do this? Why would the molecules in a fluid spontaneously order? At any given temperature, the molecules in a phase should adopt their lowest free energy arrangement, and for most fluids above their melting point, this is an isotropic liquid where the molecules are free to translate and rotate in three dimensions.

A first approximation to understanding spontaneous alignment can come from a simple macroscopic model: place a large number of matchsticks in a box and shake. You will observe that gentle shaking aligns the sticks. Shake the box hard enough and if the sticks are fairly short, they should randomize and lose their order. Of course, molecules are not rigid matchsticks, but they are continually subject to thermal fluctuations. Another way to understand spontaneous nematic ordering is to consider the entropy of the phase. The second law of thermodynamics tells us that a large system will be found in the macrostate with the greatest entropy. Our liquid crystal phase therefore should be arranged in a way that maximizes its entropy (see appendix D for more information on entropy). At first the spontaneous

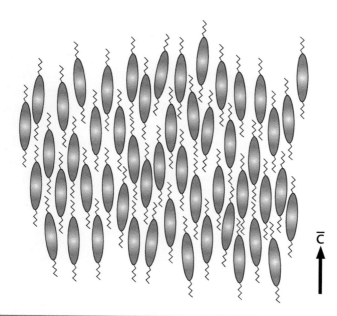

FIGURE 3.5   Schematic of the nematic liquid crystalline phase with the $\bar{c}$ director indicated.

formation of the nematic phase seems counter intuitive as a more ordered "aligned" state seemingly has lower entropy than the isotropic state in which the rod-like molecules can take any orientation. However, entropy is also affected when we increase the number of molecules per unit volume. When rod-like molecules align, the density of the phase is increased, despite the reduction in orientational freedom. Below a certain length-to-width ratio for the molecular "rods," the volume effect on entropy dominates, and it becomes entropically favorable to form an aligned nematic phase that takes up less volume.

Nematic materials are the most fluid-like of the thermotropic phases with a relatively low viscosity. In a material that exhibits several thermotropic liquid crystalline phases, the nematic phase is typically found at higher temperatures, just below the isotropic liquid phase. In Figure 3.6, we can see the molecular structure for 5CB (4-cyano-4-n-pentylbiphenyl), the first room temperature nematic material developed by George Gray at Hull University in 1972.

Although molecules in the nematic phase are, on average, aligned with the **c** director, there is a considerable degree of variation in molecular tilt with respect to this axis. Such variation can be characterized by a scalar order parameter S.

The order parameter for the nematic phase is commonly expressed as:

$$S = \frac{1}{2}\left\langle 3\cos^2\theta - 1\right\rangle \tag{3.3}$$

where θ is the angle between the molecular long axis of two neighboring molecules. Averaged over the entire system, we obtain a good measure of the degree of alignment in the nematic phase. This order parameter assumes that the molecules can be treated as non-polar rods (they don't have a head and a tail). A value of $S = 1$ represents a perfectly aligned phase (like a solid crystal), whereas $S = 0$ represents an isotropic liquid (no molecular ordering). Most nematic materials have S values between 0.3 and 0.8.

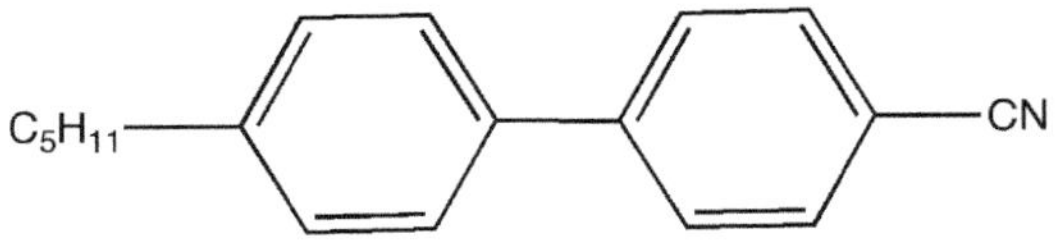

FIGURE 3.6   The molecular structure of an early and well-known nematic liquid crystal 4-cyano-4-n-pentylbiphenyl.

The nematic phase (Figure 3.5) can be identified most easily using a polarizing microscope with a thin film of the material placed between two glass plates (e.g., a standard glass microscope slide covered by a coverslip). The observer will typically see a *Schlieren texture* when the molecules are aligned parallel to the glass plates (see Figure 3.7). The nematic phase has a smooth, fluid-like appearance. Another characteristic of the nematic phase is its apparent "twinkling" or "shimmering" appearance. This effect is known as "director fluctuations" and results from continual local changes in bire-fringence as small groups of molecules fluctuate together in the sample.

The Schlieren texture is characteristic of the nematic phase and includes the topological defects we introduced in Section 3.6. The point defects seen in this phase may have two or four "brushes" (dark bands) surrounding the defect as shown in Figure 3.7. We can visualize the director field in this kind of sample because the dark and light areas of the texture correspond to different molecular orientations with respect to the axes defined by the polarizers. This concept is explained in more detail Section 3.8.2.

The nematic phase is currently the most technologically important liquid crystalline phase and can be found in most LCD screens and monitors

FIGURE 3.7 A typical planar-aligned nematic texture imaged with polarized light microscopy showing the characteristic "Schlieren texture."

in use today. Careful product optimization and development since the 1970s has resulted in commercial nematic materials with excellent properties for display applications. We will discuss how this phase is used in LCDs and other applications later in this chapter.

3.7.2 THE SMECTIC PHASES

A smectic liquid crystalline phase is characterized primarily by its layered structure. In addition to the overall alignment of the molecules with a particular $\bar{C}$ director, as in the nematic phase, smectics have an additional degree of order, the layer structure. Smectic phases are typically much more viscous than the nematic phase and occur at lower temperatures. Several different smectic-like phases have been observed and studied, each with their own unique structural features. In this section, we describe some of these.

The simplest of the smectic phases is known as smectic A (SmA) and is depicted in Figure 3.8. In this phase, the molecules tend to orient in the same direction, but in addition to this primary ordering, they arrange themselves into a layer-like structure. Thus, the phase is composed of sheets of two-dimensional (2D) fluid, stacked on top of each other. A polarized microscopy image of the smectic A phase is shown in Figure 3.9 with its characteristic defect texture.

In the smectic A phase, molecules within the layers move freely, with no defined packing arrangement and no correlation between molecules from

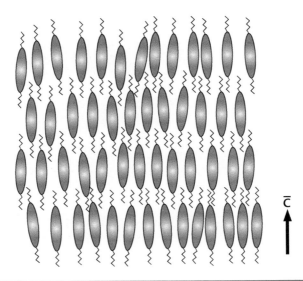

FIGURE 3.8   The layered structure of the SmA phase with the $\bar{c}$ director indicated.

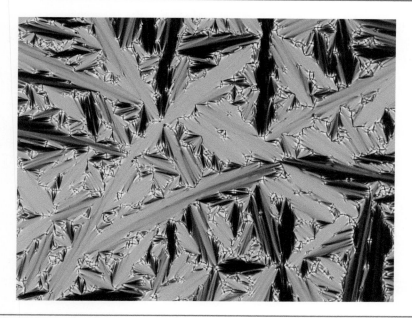

FIGURE 3.9   Focal conic texture of the smectic A phase imaged using polarized optical microscopy at 100X magnification.

layer to layer. The layer-to-layer ordering of the smectic phase can be approximated on average, to a density wave, and this model can be incorporated in a simple form of the order parameter for SmA,

$$S = \left\langle \left(\cos \frac{2\pi z}{d} \right) \left(\frac{3}{2} \cos^2 \theta - \frac{1}{2} \right) \right\rangle \qquad (3.4)$$

where z is the position of a molecule in the direction parallel to the layer normal, and d is the distance between smectic layers.

The simplest form of the smectic C (SmC) phase is a variant on the SmA phase structure. The basic layered structure is the same, but the molecules are tilted with respect to the layer normal (Figure 3.10). This tilt angle θ can vary as a function of temperature, with its maximum value at lower temperatures. At high temperatures in the phase, close to the transition to the phase above (typically SmA), the tilt angle will approach zero and can be described by the Curie-Weiss law, similar to ferromagnetism (with the transition temperature between phases analogous to the critical temperature T_c),

$$\theta = \left(T - T_c \right)^{-\gamma} \qquad (3.5)$$

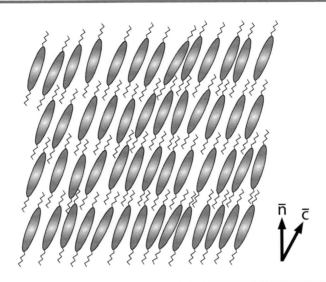

FIGURE 3.10 The SmC phase. The $\bar{c}$ director and the layer normal, $\bar{n}$ are indicated.

The SmA to SmC phase transition is a second-order phase transition, there is no discontinuity in order parameter at the transition point (i.e., the tilt varies smoothly, tending to zero at the transition to SmA).

3.7.3 CHIRALITY IN LIQUID CRYSTALS

Chirality is an essential property of several interesting liquid crystalline phases, so it is important to introduce this concept here. Up to this point, we have modeled the liquid crystal molecule as a rod-like axisymmetric object. In fact, many molecules are not perfectly rod-like, but have a certain "handedness" or "chirality" that restricts how they can pack together. To help visualize this, think about packing some springs together. Two springs or screw threads can pack fairly well with a slight twist by slotting together their threads. However, if you replace one of the springs with one of an opposite handedness, now packing them becomes very unfavorable and the screw threads will not slot together. Another example of chiral packing can be seen in your own hands. Try to place them on top of each other, and you quickly see that the only way to pack a large number of hands efficiently will be to pack all left hands or all right hands. Objects of opposite "handedness" cannot be stacked on top of each other in a space-efficient manner (i.e., a chiral structure cannot be superimposed on its mirror image).

If multiple springs or other chiral objects are packed together closely, you quickly notice that an overall "twist" appears to develop. Each spring, when packed next to its neighbor, will be rotated by a small angle as the threads slot into each other. They do not align perfectly like smooth rods, and this effect gives rise to "bulk chirality." So, individual molecules with a chiral nature can give rise to chiral phases. Chiral phases have an overall twist that produces interesting optical properties.

3.7.4 THE CHOLESTERIC PHASE

The best known of the chiral liquid crystal phases is the *cholesteric phase* or chiral nematic (*N**—here an asterisk is used to indicate a chiral phase). The cholesteric phase was the first liquid crystal to be discovered by Reinitzer in 1888. Reinitzer observed pure cholesterol benzoate under the microscope and noticed two apparent melting points: the solid crystal form first melted into a phase that is now known as the cholesteric phase, and then he observed a second melting point at which the cholesteric phase melted into an isotropic liquid phase. Cholesterol (Figure 3.11) is a chiral molecule with a relatively rigid central core. Optical microscope images of a planar-aligned cholesteric phase are shown in Figures 3.12 and 3.13.

The structure of the cholesteric phase is represented by the schematic in Figure 3.14. Imagine starting on the top and moving down through a thick slab of cholesteric phase. The molecular orientation varies continuously by rotating in-plane through the slab with molecules on the top having a slightly different orientation to molecules located directly below. The cholesteric phase is not layered like a smectic phase and is more akin to the nematic phase in terms of molecular ordering and viscosity. Hence, the alternative name *chiral nematic (or N**). We can define a pitch, *P* for the cholesteric phase as the distance through which the molecule makes a complete in-plane twist (360°). Typical values for the pitch in many materials are around 200–700 nm. When white, unpolarized light is incident on a slab of cholesteric, the material will reflect circularly polarized light at a wavelength that

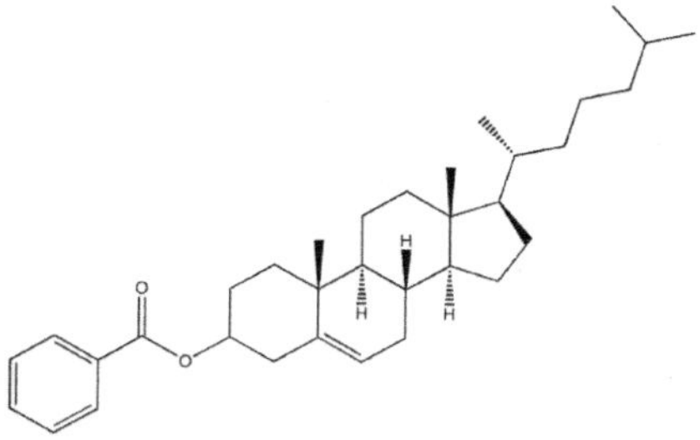

FIGURE 3.11 Cholesterol derivatives were the first liquid crystals investigated. Cholesterol benzoate shown here is a chiral molecule and forms a cholesteric phase.

FIGURE 3.12 Polarized microscopic image of a planar-aligned cholesteric phase ("Grandjean" or "oily streak" texture).

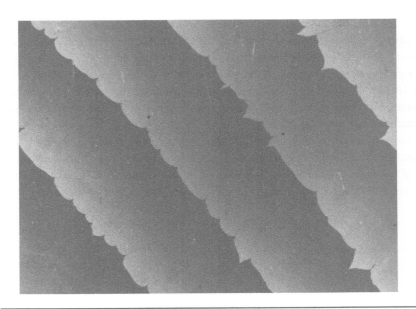

FIGURE 3.13 Polarized optical microscopic image of cholesteric liquid crystal in a Grandjean-Cano wedge cell geometry. In this geometry, a thin film of cholesteric liquid crystal of gradually increasing thickness is prepared between non-parallel glass plates. The cholesteric helix stretches to accommodate the increasing film thickness until at the sharp steps an additional 180° twist takes place. The repeated expansion of the helix is revealed by the reflection colors seen in this image. (Courtesy of Andrea Rodarte.)

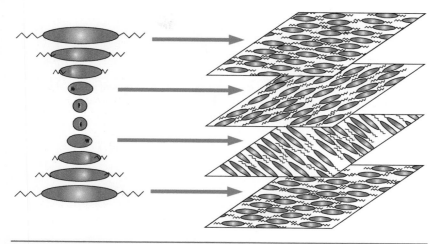

FIGURE 3.14   Schematics of the cholesteric phase showing the "twist" in molecular orientation through the sample.

matches the pitch *P*. This optical effect is particularly useful for some applications because cholesteric films can be designed to reflect specific wavelengths of light. Another important and widely exploited property of the cholesteric phase is its sensitivity to changes in temperature. As temperature is increased and the cholesteric-to-isotropic transition point is approached, the pitch of the phase decreases significantly. This effect is exploited in temperature sensors, and we will take a look at this application in a little more detail in Section 3.9.4.

3.7.5   THE CHIRAL SMECTIC PHASES

In Section 3.7.2, we described the SmA and SmC phases, in which achiral rod-like molecules pack together into layered phases. Certain liquid crystal molecules with the potential to form these layered phases are also chiral (Figure 3.15), and this property can lead to the formation of *chiral* smectic phases. The chiral smectic C phase (SmC*) is the most well-known. This phase is essentially just a chiral version of SmC. The molecules are arranged in a layered structure with a tilt angle θ with respect to the layer normal. In addition to this positional ordering, however, from layer to layer, the azimuthal angle φ of the molecule precesses slightly (the geometry of a single molecule is defined in Figure 3.16). Similar to the cholesteric phase, we can also define a pitch *P* in chiral smectics. This corresponds to the thickness of material through which φ varies by 360°. This combination of chirality and molecular tilt led to the discovery of *ferroelectricity* in liquid crystals, as predicted in 1975 by Robert Meyer.[3] Smectic molecules may exhibit electric polarization (i.e., they

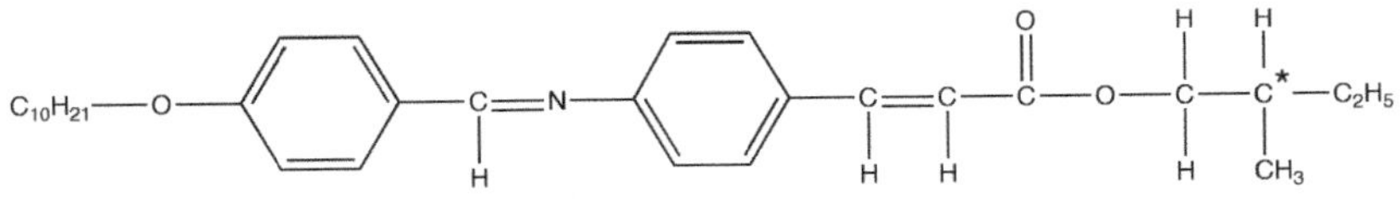

C₁₀H₂₁—O— ... —C=N— ... —C=C—C—O—C—C*—C₂H₅

FIGURE 3.15   An example of an early chiral smectic liquid crystal material *p*-(*n*-decyloxybenzylidene)-*p*-amino-(2-methyl-butyl) cinnamate (DOBAMBC).

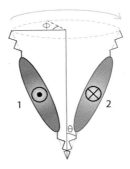

FIGURE 3.16   Diagram illustrating the geometry of ferroelectric switching between azimuthal positions 1 and 2 in the SmC* phase.

are electric dipoles and can respond to an electric field—more on this later in the chapter). Ordinarily in a chiral phase, the net polarization of the individual molecules will be canceled out by the presence of the helix. It is, however, possible to "unwind" this helix by forming a thin layer of smectic (~5 μm) between glass plates. This configuration produces a net polarization, and as a result molecular "switching" between the two states (1 and 2) can take place, as demonstrated in Figure 3.16. This switching can only occur because ferroelectric liquid crystals are composed of molecules with a permanent electric dipole moment μ. In the presence of an electric field, the electric dipole will align with the field. If an alternating E-field is applied to the material, the molecules will respond and continuously move to align with the changing E-field.

Figure 3.16 shows a common representation of the ferroelectric switching mechanism. Initially, a liquid crystal molecule in the smectic layer is depicted as lying on the side of an imaginary cone at a tilt angle θ. The electric dipole moment vector μ is indicated as coming out of the page. If an E-field is applied to the molecule at position 1 into the page, then for μ to align with the field, the molecule must reorient, swinging around the outside of the cone to position 2. This ferroelectric switching mechanism is also known as the *Goldstone mode* and has been used in various ferroelectric liquid crystal technologies.

This liquid crystal device geometry was first applied in 1980 in the surface-stabilized ferroelectric liquid crystal display[4] and provided much faster switching times than the nematic devices of the time (<0.1 ms); however, the main drawback of this smectic device is the stability of liquid crystal alignment within the pixels. Nematics are very fluid-like, and after a deformation, they rapidly revert to their previous uniform state of alignment (think about what happens when you press on your laptop screen). Smectics are much more viscous and unfortunately do not self-repair when deformed.

3.7.6 OTHER CHIRAL SMECTIC PHASES

Chiral smectic liquid crystal materials are best known for their ferroelectric properties, and after the discovery of these materials a considerable amount of attention was dedicated to their development for display applications. In the late 1980s, researchers began to observe some surprising switching behavior in SmC* materials, and it was found that certain materials exhibited a new liquid crystal phase with antiferroelectric-like properties, the SmC$_A$* phase.[5] In this so-called antiferroelectric phase (Figures 3.17 and 3.18), all molecules within a particular smectic layer are tilted and have the same azimuthal orientation, but this azimuthal angle varies by a 180° rotation from layer to layer, while still exhibiting the gradual chiral twist seen in the SmC* phase and the cholesteric phase (chiral nematic). The antiferroelectric phase is particularly interesting because it can be forced to adopt three different states by applying electric fields to the material as shown in Figure 3.17, thus providing the possibility of three different switching states in a device instead of the two states used in nematic or ferroelectric displays.

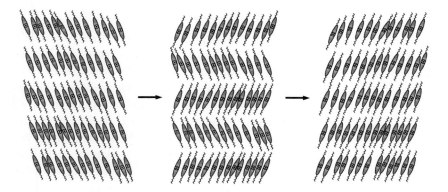

FIGURE 3.17 The structure of the antiferroelectric liquid crystal phase (center) and electric-field-induced switching states.

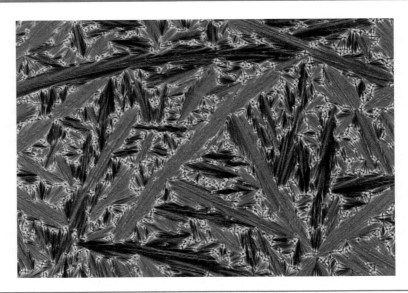

FIGURE 3.18 Polarized microscope image of a planar texture in the antiferro-electric phase.

Since the discovery of the antiferroelectric phase, several additional chiral smectic phases have been observed and studied in detail, most notably the three-layer (Figure 3.19)[6] and four-layer intermediate phases (also known as ferrielectric), SmC^*_{FI1} and SmC^*_{FI2}. The structures of these phases are defined by repeated variations in azimuthal angle from layer to layer and have shown promise for the development of interesting new liquid crystal devices with many different switching states, but technological applications have yet to be realized.

3.7.7 THE BENT-CORE (BANANA) PHASES

Previously in this chapter, we discussed the properties of calamitic (rod-shaped) liquid crystal molecules and how the packing of their rigid cores can lead to liquid crystal phases. There is another class of liquid crystal molecules with a different molecular shape that has generated some considerable attention in recent years. Bent-core liquid crystals, also known as "banana" molecules or "bow-shaped" molecules, have a rigid core with a bend near their center. Figure 3.20 shows an example of one of these molecules. The first bent-core liquid crystal materials were synthesized as long ago as 1929 by Vorlander, although their interesting properties were not appreciated at the time. In fact, it took until the early 1990s for the study of bent-core liquid crystals to really begin.

As you can imagine, when bent-core molecules pack together, a variety of different phases can form, distinct from the more conventional calamitic

FIGURE 3.19   Polarized microscope image of the homeotropic texture of the three-layer intermediate (also known as the ferrielectric phase). The three-layer phase is optically indistinguishable from the four-layer phase, but the two phases can be clearly differentiated using the resonant x-ray scattering technique[6].

phases formed from simple rod-like molecules. These different phases result from the additional complexity added to packing "bent" molecules instead of "rod-like" ones. There are eight main bent-core phases (B1–B8). In fact, several variants on these core phases have been proposed in recent years, and the complete family of bent-core phases is still under investigation.

Bent-core liquid crystal materials are of particular interest because they exhibit some novel properties, including fast responses to electric fields and biaxiality. An in-depth discussion of the bent-core liquid crystal phases is beyond the scope of this book and remains an active area of research; for more details, see the further reading suggestions at the end of this chapter.

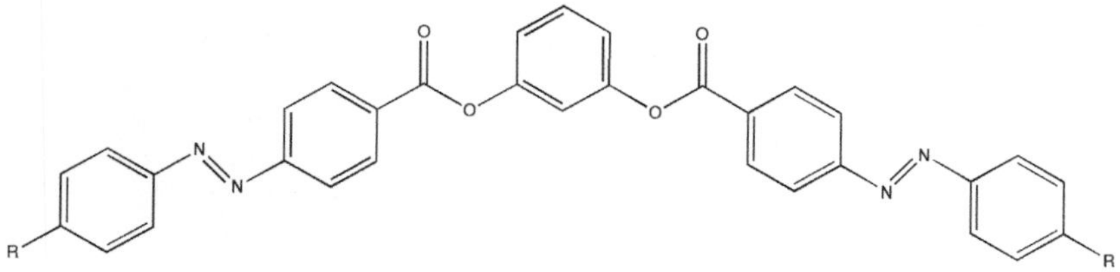

FIGURE 3.20   Molecular structure of a historically important bent-core liquid crystal molecule synthesized by Vorlander.

3.7.8 DISCOTIC PHASES

The *discotic* phases represent another class of liquid crystal materials that have received a lot of research attention. As the name implies, these phases form from disk-shaped molecules and a material showing this behavior was first identified and studied in 1977 by Chandrasekhar. Examples of some typical discotic molecular structures are shown in Figure 3.21. There are two main classes of discotics, the nematic and the columnar discotic phases; schematics of these phases can be seen in Figure 3.22. The nematic discotic phase is analogous to the calamitic nematic phase. Molecules are aligned in roughly the same direction (in this case, defined by the normal to the plane of the disk), but exhibit fluid-like properties throughout the phase. There is no long-range order, and molecules can translate freely throughout the bulk. The columnar phase forms when the disk-like molecules become stacked on top of each other in columns. Ordering within an individual column

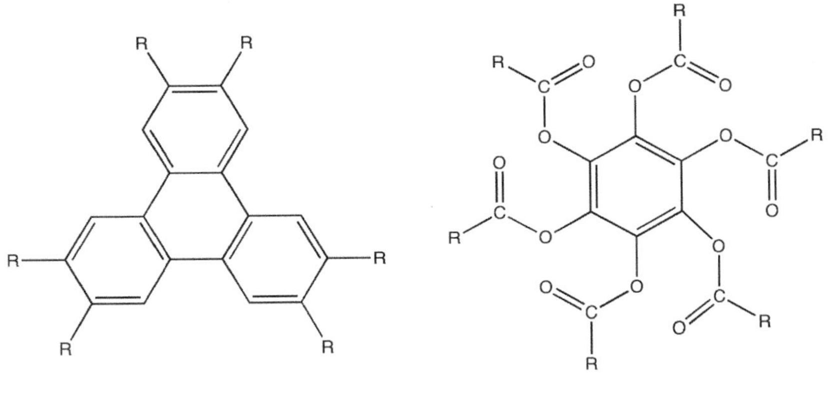

FIGURE 3.21 Molecular structures of some early discotic liquid crystal materials.

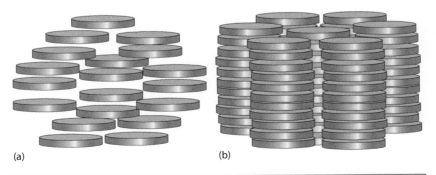

(a)

(b)

FIGURE 3.22 Schematic of the (a) nematic discotic and (b) hexagonal columnar discotic phases.

is fluid-like, but the columns themselves may also be arranged in a lattice arrangement. The in-plane packing of the columns is most commonly hexagonal, although other variants have been identified and can result from varying the molecular shape.

3.8 EXPERIMENTAL TECHNIQUES

3.8.1 DEFORMING LIQUID CRYSTALS

Liquid crystals are fluids, but they can also be deformed like an elastic solid. This might seem surprising at first, but if we consider that the relaxed (equilibrium) state of the fluid is molecular alignment, then forcing the molecules away from this state by some external mechanism should cost energy and can thus be considered a deformation. The molecular orientation (i.e., the director) in liquid crystal materials can be deformed under the influence of mechanical shear, using surfaces that promote a particular molecular anchoring or by the action of an external electric or magnetic field. Deformations to the director in the nematic phase can be classified in three different ways (i.e., there are only three ways in which the nematic ordering can be deformed). These are known as splay, twist, and bend. (Note that in this section, we consider only the nematic phase for simplicity.) These three different modes of deformation are demonstrated in Figure 3.23.

After a deformation force is removed from the liquid crystal material, it will tend to relax back to its equilibrium state. This behavior is analogous to elasticity in continuous solid, and it is possible to describe the deformation of liquid crystals in terms of continuum elastic theory. When we describe the elastic properties of a solid, we think of Hooke's law,

$$F = -k\Delta x \qquad (3.6)$$

Where F is the restoring force, k represents an elastic constant for the deformable material, and Δx represents the deformation from equilibrium.

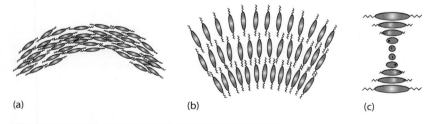

(a)   (b)   (c)

FIGURE 3.23 The three different modes of liquid crystal deformation: (a) bend, (b) splay, and (c) twist.

The most familiar realization of Equation (3.6) is the spring and the energy associated with a simple elastic deformation is,

$$E = \frac{1}{2}k(\Delta x)^2 \tag{3.7}$$

It is interesting to think about how elastic theory can be applied to a fluid, most simple liquids do not have this property. Liquid crystals are different, however, their equilibrium state is anisotropic (neighboring molecules align spontaneously). By locally disturbing this alignment, we deform the fluid from equilibrium, an action which costs energy. Continuum elastic theory was extended successfully to liquid crystals by Frank in 1958. The following expression (Equation 3.8) represents the general form in which the free energy F of a deformed liquid crystal medium is usually expressed:

$$F = \frac{1}{2}K_{11}\left(\nabla \cdot \hat{n}\right)^2 + \frac{1}{2}K_{22}\left(\hat{n} \cdot \nabla \times \hat{n}\right)^2 + \frac{1}{2}K_{33}\left(\hat{n} \times \nabla \times \hat{n}\right)^2 \tag{3.8}$$

This expression, while appearing very complex, can be simply broken down into three parts, each representing a different mode of deformation, or curvature of the director field n.

$$\text{Splay } \frac{1}{2}K_{11}\left(\nabla \cdot \hat{n}\right)^2 \tag{3.9}$$

$$\text{Twist } \frac{1}{2}K_{22}\left(\hat{n} \cdot \nabla \times \hat{n}\right)^2 \tag{3.10}$$

$$\text{Bend } \frac{1}{2}K_{33}\left(\hat{n} \times \nabla \times \hat{n}\right)^2 \tag{3.11}$$

The fluid-like nature of the nematic phase allows these deformations to take place with relatively little energy cost (compared to a crystalline solid). In these equations, the constants K_{11}, K_{22}, and K_{33} are known as the *Frank elastic constants* and typically take a value around 10^{-11} N/m. Measurement of the elastic constants is often carried out to characterize the "switching" capabilities of liquid crystal materials (i.e., how well they will perform in a device under the application of an electric field). Elastic effects in liquid crystals are also important in the field of nanocomposites—whereby small particles are dispersed in either a liquid crystal or a polymer phase to provided added functionality. Imagine trying to insert a spherical particle with homeotropic surface anchoring into an aligned nematic phase—can you sketch what the liquid crystal molecules would do to incorporate the particle with minimal elastic energy cost?

3.8.2   POLARIZED OPTICAL MICROSCOPY

As we have learned, liquid crystal materials may exhibit birefringence (see Section 3.5), so *polarized optical microscopy* is an ideal technique for visualizing liquid crystal textures. The liquid crystal is prepared as a thin film (~2–20 µm thick) between glass plates. By observing the interesting defect textures that may form in each phase, it is often possible to make accurate phase identifications by microscopy alone, even though the microscope is unable to resolve the actual molecular packing structure. Figure 3.24 shows the geometry of a typical polarized microscope.

For phase identification, it is useful to be familiar with the different birefringence textures each phase can exhibit. These textures depend on the spatial variation in molecular alignment with respect to the optical axis and the polarizers. To illustrate this, we can consider the Schlieren texture of

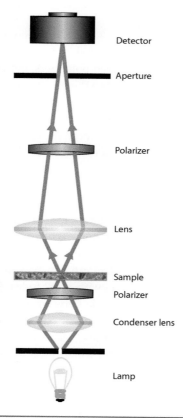

Detector

Aperture

Polarizer

Lens

Sample

Polarizer

Condenser lens

Lamp

FIGURE 3.24   Schematic showing the operation of the polarizing optical microscope. Two polarizers on either side of the sample are crossed so differences in birefringence in the sample may be observed.

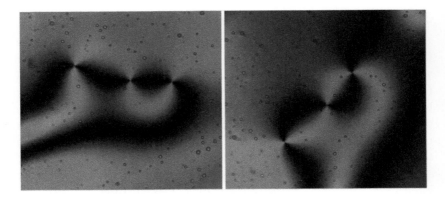

FIGURE 3.25 Two different polarized optical microscopic images of the same "brush" defect in the nematic phase taken at two different arbitrary angles 45° apart. Notice how the dark areas are different in the two images as different areas of the sample become aligned with the polarizer direction.

the nematic phase shown in Figure 3.25. In this figure, two different views of the same nematic defects are presented. In both images, the liquid crystal is shown illuminated between crossed polarizers, but the right-hand image has been rotated by 45° with respect to the one on the left. So, why are some areas dark in one view and then light in the other? It turns out that the intensity of the transmitted light through the system depends on the orientation of the liquid crystal director with respect to the crossed polarizers. Is it possible to find the topological charge of these defects from the images?

In Figure 3.26, we can see a more detailed view of the liquid crystal slab and the polarizers. To understand how molecular orientation is connected to transmitted light intensity, we start by considering unpolarized incident monochromatic light incident on the system shown in Figure 3.26. When light is incident on the first polarizer, oriented in the y direction, it will become linearly polarized, with an intensity I_0. This polarized light is incident on the liquid crystal slab and can be resolved into two waves polarized in orthogonal directions, with amplitudes $A\sin\theta$ and $A\cos\theta$. As indicated in Figure 3.26, θ is the angle between the liquid crystal director and the polarizer (P1) direction. We can treat these orthogonal components independently as they propagate through the light crystal slab. The component polarized parallel to the director will exit the slab with an amplitude of:

$$A\cos\theta \, \cos\left(\omega t - \frac{2\pi}{\lambda}n_e d\right) \tag{3.12}$$

and the component polarized perpendicular to the director will have an amplitude of

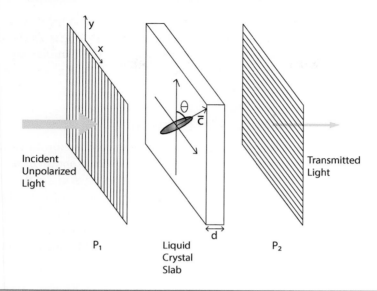

FIGURE 3.26 The geometry of a liquid crystal slab rotted between crossed polarizers.

$$A\sin\theta\,\cos\left(\omega t - \frac{2\pi}{\lambda}n_o d\right) \qquad (3.13)$$

Finally, after passing through the analyzer (P2), aligned in the x direction, the two transmitted wave components will have amplitudes equal to:

$$-A\cos\theta\sin\theta\cos\left(\omega t - \frac{2\pi}{\lambda}n_e d\right) \qquad (3.14)$$

and

$$A\cos\theta\sin\theta\,\cos\left(\omega t - \frac{2\pi}{\lambda}n_o d\right) \qquad (3.15)$$

The resultant intensity I can be found by squaring the sum of these amplitudes. Then, trigonometric identities will simplify our result to,

$$I = I_o \sin^2 2\theta \sin^2\left(\frac{\pi d}{\lambda}\Delta n\right) \qquad (3.16)$$

where Δn is the birefringence ($n_e - n_o$). From 3.16, we can clearly see that as the liquid crystal sample is rotated between the polarizers by angle θ, the transmitted intensity will vary as $\sin^2 2\theta$, from 0 (when the director is aligned with either polarizer), to a maximum value at 45°. So, as the liquid crystal director direction varies spatially across the material (e.g., in Figure 3.25), transmitted intensity also varies. The image is a kind of map representing molecular orientation!

3.8.3 ELECTRO-OPTICAL MEASUREMENTS

Polarized microscopy is particularly powerful when combined with electrical measurements on liquid crystal films (known as electro-optical measurements). From microscopic observations of the birefringence in a material as a function of different applied fields, changes in molecular orientation can be inferred, and it is possible to deduce the mechanisms by which the liquid crystal molecules respond to electric or magnetic fields. It is common to look at the effects of electric fields on liquid crystal materials. Although magnetic fields will also result in molecular reorientation, the effect is much weaker and therefore has not been commercially useful.

3.8.4 THE DIELECTRIC PROPERTIES OF LIQUID CRYSTALS

The dielectric constant of an insulating material ε, also known as the permittivity, can be defined as the extent to which a material will become electrically polarized in the presence of an electric field. In an isotropic material, we define a single dielectric constant. However, the anisotropic nature of liquid crystal materials, such as the nematic phase, means that the direction of the electric field with respect to the liquid crystal director is important. In this case, we can define two different dielectric constants, $\varepsilon_\parallel$ and $\varepsilon_\perp$. $\varepsilon_\parallel$ represents the ability of the molecule to polarize along the long axis and $\varepsilon_\perp$ to polarize along the short axis. The existence of these two different dielectric constants means that we can define a new quantity, the *dielectric anisotropy*,

$$\Delta\varepsilon = \varepsilon_\parallel - \varepsilon_\perp \tag{3.17}$$

The dielectric anisotropy, $\Delta\varepsilon$ can be either positive or negative, with positive $\Delta\varepsilon$ materials orienting with their long axis parallel to the E-field and negative $\Delta\varepsilon$ materials orienting with their short axis parallel to the E-field. Practically, the magnitude of $\Delta\varepsilon$ represents the ease with which the molecules will respond to an applied field; the larger the value of $\Delta\varepsilon$, the easier it is to reorient the liquid crystal molecules.

The dielectric constants for a liquid crystal material can be determined by treating the liquid crystal cell as a parallel plate capacitor filled with a liquid crystal as the dielectric. The capacitance of this cell is given by:

$$C = \frac{\varepsilon_0 \varepsilon A}{d} \qquad (3.18)$$

where A is the area of the capacitor, and d is the separation of the plates. By carrying out two experiments, one in which the liquid crystal is oriented homeotropically and another with planar alignment, $\varepsilon_\parallel$ and $\varepsilon_\perp$ can be determined independently.

3.8.5   THE FRÈEDERICKSZ TRANSITION AND MEASUREMENT OF THE ELASTIC CONSTANTS

In general, when a liquid crystal molecule is subject to an electric field the molecule will tend to become electrically polarized. This polarization creates a dipole in the molecule that will tend to align with the electric field. In the bulk nematic phase, aligned molecules will reorient together in response to the applied field. This bulk reorientation of the nematic director field is known as the *Frèedericksz transition.*

The deformation most commonly applied in a Frèedericksz transition is the splay deformation, and this geometry can be seen in Figure 3.27 for just a few liquid crystal molecules in a planar geometry. We can use the Frank

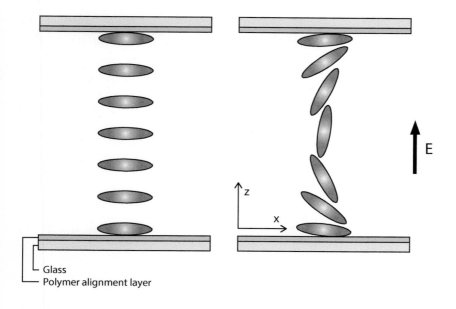

FIGURE 3.27   The response of a planar aligned liquid crystal with a positive dielectric anisotropy to the application of a perpendicular electric field, E. A single column of molecules is shown for clarity.

elastic constant for splay, K_{11}, to quantify the energy it takes to deform the liquid crystal in this way; further on in this chapter, we will see how this field-induced transition is applied to liquid crystal displays. A more detailed description of the Fréedericksz transition is also given in *Fundamentals of Liquid Crystal Devices*, by Deng-Ke Yang and Shin-Tson Wu,[7] in which the following equations are derived in detail. In a liquid crystal device, there are two competing responses when an electric field is applied. On the one hand, the molecular dipoles will respond to the electric field and reorient. A molecule with positive dielectric anisotropy will align with the field; however, there is an elastic energy cost to this response in a real device. We can assume that the molecules close to the surfaces are strongly anchored and will not reorient; therefore, in the case shown in Figure 3.27, a splay deformation is required to accommodate the reorientation. Here, we are just considering a system with splay deformation only, although in general there could also be twist or bend deformations with the application of an arbitrary field.

When an electric field is applied to a molecule oriented at an angle θ to the E-field (Figure 3.28), the change in free energy as the molecule rotates to align with the E-field is given approximately by,

$$f_e = -\frac{1}{2}\varepsilon_0\Delta\varepsilon E^2\sin^2\theta \qquad (3.19)$$

that is, a field-induced molecular reorientation will lower the free energy of the system. The angle θ is defined as the angle the molecular director makes with the E-field axis. The elastic properties of the liquid crystal phase will act to oppose this motion with an increase in free energy (for splay only) of:

$$f_{\text{splay}} = \frac{1}{2}K_{11}\left(\nabla.\hat{n}\right)^2 \qquad (3.20)$$

If we define the director **n** as:

$$\hat{n} = \cos\theta\,(z)\hat{x} + \sin\theta\,(z)\hat{z} \qquad (3.21)$$

which is a vector in the x–z plane varying in direction as a function of z, then we can write:

$$f_{\text{splay}} = \frac{1}{2}K_{11}\left(\frac{\partial\theta}{\partial z}\cos\theta\right)^2 \qquad (3.22)$$

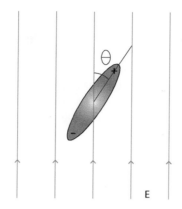

FIGURE 3.28   The electric dipole in a liquid crystal molecule with positive dielectric anisotropy rotates through angle θ to align with an applied electric field, E.

Balancing these free-energy expressions and integrating over the width of the device d we obtain the total free energy per unit area:

$$F = \int_0^d \left[-\frac{1}{2}\varepsilon_0 \Delta\varepsilon E^2 \sin^2\theta + \frac{1}{2}K_{11}\left(\frac{\partial\theta}{\partial z}\cos\theta\right)^2 \right] dz \qquad (3.23)$$

In the strong surface anchoring limit, the threshold electric field for a splay Fréedericksz transition, as shown in Figure 3.25, is given by the following equation[7]:

$$E_c = \frac{\pi}{d}\sqrt{\frac{K_{11}}{\varepsilon_0 \Delta\varepsilon}} \qquad (3.24)$$

where K_{11} is the splay elastic constant, E_c is the threshold electric field, and d is the thickness of the liquid crystal material. Notice from this equation that the threshold field is inversely proportional to the device thickness. This result is expected because in a thin device the effects of surface anchoring will mean that the elastic energy required for the deformation will be larger.

Experimental measurement of the threshold voltage for the Fréedericksz transition provides us with a method for determining the Frank elastic constants. The constants K_{11}, K_{22}, and K_{33} each represent, respectively, the ease with which splay, twist, and bend deformations can be made. Measurements of the elastic constants are carried out by preparing liquid crystal cells in different geometries and determining the threshold field for the Fréedericksz transition. If we can prepare liquid crystal cells each with just one of the three deformations and find the threshold voltage for the Fréedericksz transition, then each of the elastic constants can be determined independently. This measurement of the Frank elastic constants is particularly important when assessing the ability of a material to reorient in an electric field (i.e., for a nematic device, such as a switchable pixel).

3.8.6   X-RAY DIFFRACTION

X-ray diffraction is an experimental technique that can be used to deduce the molecular packing structure in a liquid crystal phase. In the same way that diffraction methods are used to solve the structures of different solid crystals, we can gain structural information on soft materials such as liquid crystals using this technique. X-ray scattering measurements on liquid crystals can be carried out using either a lab-based x-ray generator or a more intense x-ray source, such as a synchrotron facility. The scattering intensity of x-rays from liquid crystal materials, as with most soft materials, is significantly lower than that from hard crystalline solids. This difference means that a much more intense x-ray source is often needed to obtain useful results. Another difference in scattering from liquid crystal materials comes from the fact that the phases typically have short-range positional order. The diffraction patterns are quite different from those observed for conventional crystals. No sharp diffraction peaks are observed; instead, more diffuse features identify the different phases. The scattering patterns shown in this section correspond to aligned liquid crystal samples in which the molecular axis has been controlled throughout the sample. This is the most reliable way to determine phase from scattering patterns. Unaligned measurements (confusingly sometimes referred to as "powder" patterns even though we are still in the liquid crystal state) are simply scattering measurements performed on liquid crystal with no preferred orientation direction (in a similar way to taking solid crystal and crushing it into a powder). They are composed of a mosaic of different randomly oriented liquid crystal domains. This effect produces a scattering pattern smeared out into diffuse rings.

In a typical x-ray scattering experiment on a liquid crystal material, the sample is loaded into a holding cell able to contain the material in a liquid-like state. This may be a thin film sealed between thin glass plates (the glass should be as thin as possible to reduce x-ray beam attenuation), a glass or quartz capillary, or more simply a sealed capsule prepared from either Mylar or Kapton film (two polymer films with low x-ray absorption). The sample capsule may be mounted on a temperature-controlled stage in the path of the x-ray beam. In a transmission experiment, the x-ray beam passes through the sample, and scattering x-rays are recorded on a detector. This can be an area detector, such as a CCD (charge-coupled device) or an image plate. For higher resolution, a point detector can be used. Figure 3.29 depicts a typical experimental setup for these experiments.

Figure 3.30 shows a cartoon representation of some typical scattering patterns for the smectic and nematic liquid crystal phases. The smectic phase has a layered structure with an approximately sinusoidal density

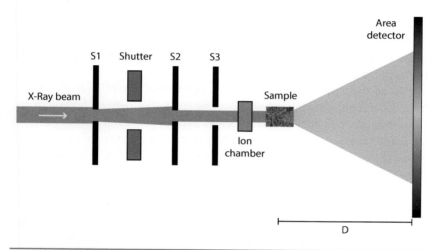

FIGURE 3.29   Schematic of a typical transmission x-ray scattering beamline. The beam size is defined by slits S1 and S2; S3 is a guard slit spaced more widely to reduce parasitic scatter. A simple intensity detector such as the ion chamber shown is used to monitor beam intensity at the sample; then after the sample, a distance D away, scattered x-rays are detected. Area detectors such as a charge-coupled device are commonly used; however, higher-resolution data can be obtained using a movable point detector.

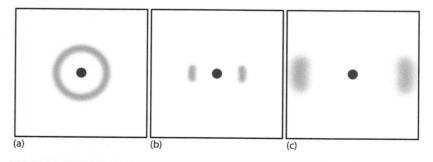

FIGURE 3.30   These cartoons illustrate the different scattering patterns that may be observed in an x-ray diffraction experiment on a liquid crystal material for (a) an unaligned smectic phase, (b) an aligned smectic A phase, and (c) an aligned nematic phase.

distribution from layer to layer. This 1D ordered structure typically produces a single scattering peak (using a high-intensity source, a second order can be observed, but this is typically very weak), and in an unaligned sample, a ring of scattered x-rays is observed on the area detector (Figure 3.30a). The simple Bragg's law:

$$n\lambda = 2d \sin \theta \tag{3.25}$$

can be used to deduce the *smectic layer spacing d*, where λ is the x-ray wavelength, n is the order of the Bragg peak (an integer), and θ is the scattering angle, which can be measured if the position of the peak on the detector and the sample-to-detector distance are known.

In an unaligned sample, the different smectic phases typically cannot be independently distinguished from the scattering pattern (although the phase boundaries between the SmA and SmC phases can be determined in the same sample by plotting the layer spacing as a function of temperature across the transition). Much more detailed information can be gleaned using aligned samples. An example of how scattering from an aligned SmA phase would look is seen in Figure 3.30b. Additional complications to the experiment can arise, however, for aligned samples. For example, in smectics, it is desirable to align the smectic layers parallel to confining glass plates. However, at the SmA to SmC transition, a change in layer spacing occurs as the molecules tilt within the layers, causing the layers to buckle in the middle of the device, producing a chevron structure. This produces two sets of layers with different orientations in the device. In this case, the aligned x-ray experiment must be carefully designed and the sample oriented correctly to scatter from the layers. Figure 3.30c shows an example scattering pattern obtained from an aligned nematic liquid crystal phase. In this phase, the molecules lack positional order; however, there is a still a characteristic *correlation length* between molecules. This distance is consistent enough that a diffuse scattering peak can be observed corresponding to the approximate end-to-end intermolecular distance. A second diffuse peak (not shown in the figure) can also be detected at a wider angle corresponding to the side-to-side correlations between nematic molecules.

3.8.7 DIFFERENTIAL SCANNING CALORIMETRY

Calorimetry is a thermal characterization technique often used to identify and study phase transitions. The technique is particularly valuable in liquid crystal materials as it assists in the accurate measurement of transition temperatures for different phases. When a material is heated, the temperature of that material varies linearly as a function of the thermal energy applied. This relationship is quantified by the heat capacity, c. The heat flow ΔQ into a material is equal to:

$$\Delta Q = cm \, \Delta T \qquad (3.26)$$

where m is the mass of the material, and ΔT is a change in temperature.

The differential scanning calorimeter (DSC) is an instrument that heats a solid or liquid sample in parallel with a reference (usually an empty sample pan). The instrument is calibrated such that the temperature of the two chambers is kept the same as they are heated at a fixed rate (°/min) (Figure 3.31). A heat flow curve is measured by the instrument as the material is heated or cooled. For a material that exhibits no thermal phase transitions, this heat flow curve will be linear with a slope indicative of heat capacity of the material. At a phase transition (e.g., melting a crystalline solid into a liquid crystal phase), a discontinuous change in heat flow into the sample occurs to maintain the temperature rate. Extra heat is either required for the transition to take place or heat is released by the transition. This change will be measured as a deviation from the baseline heat flow curve, as a peak (for an endothermic process), a trough (for an exothermic process), or perhaps a more subtle change, such as a change in slope (typical for the polymer glass transition).

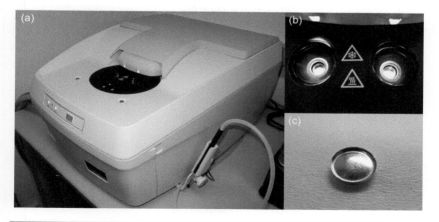

FIGURE 3.31   A high-sensitivity differential scanning calorimeter typically used for liquid crystal materials is shown in (a) and a close-up picture of the two furnaces in (b). Picture (c) shows an aluminum sample pan used to encapsulate the sample. These sample pans hold a few milligrams of material.

Different liquid crystal phase transitions will be more or less difficult to detect using DSC. If the transition is first order, meaning that the order parameter is discontinuous across the phase boundary, a significant latent heat will be measurable, and usually a clear peak can be observed. An example of a first-order phase transition in liquid crystals would be the crystalline-to-smectic or -nematic phase. The nematic-to-isotropic phase transition is weakly first order, with a relatively small but discontinuous change in enthalpy. Some liquid crystal phase transitions are much more difficult to observe using DSC, and even though there is a discontinuous structural change, the enthalpy difference can be very difficult to detect, with no distinct peak and/or just a subtle discontinuity in heat capacity at the transition. Figure 3.32 shows an example of DSC data for a smectic liquid crystal. In these data, we can observe two significant peaks at 32.16°C and 39.56°C. The transition point can be measured as either the maximum or the onset of this peak, and the area under the curve will yield the enthalpy of the transition ΔH. This is equal to the energy required or given off when one mole of material changes phase where,

$$\Delta H = \Delta U + P\Delta V \tag{3.27}$$

and U is the internal energy of the material. The latent heat of the transition between phases A and B (L_{AB}) is equal to the difference in enthalpy between the two states,

$$L_{AB} = H_B - H_A \tag{3.28}$$

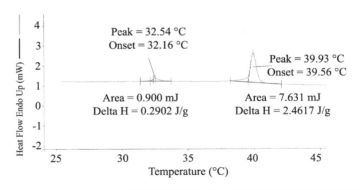

FIGURE 3.32 An example of DSC data for a thermotropic liquid crystal (8CB) measured on a heating run of 20°/min. Two peaks can be observed. The first peak with an onset at 32.16° is the smectic-A-to-nematic phase transition, and the second peak at 39.56° is the nematic-to-isotropic liquid phase transition. Delta H is the enthalpy of the transitions.

3.9 APPLICATIONS OF LIQUID CRYSTALS

3.9.1 LIQUID CRYSTAL DISPLAYS

The most well-known application of a liquid crystal is of course, the liquid crystal display. After extensive development, this technology has come to almost completely displace the cathode ray tube (CRT) as the dominant display on the market and in our homes and offices. LCD computer monitors (Figure 3.33) and televisions bigger than 50 inches are now widely available and affordable for the consumer.

Almost all LCDs on the market today contain nematic liquid crystal as their active medium. These nematic materials are composed of mixtures of several liquid crystal compounds and have been carefully designed to exhibit

FIGURE 3.33 A liquid crystal display displaying a photograph of a liquid crystal (a) with close-ups of the screen, (b) individual pixels are visible, and (c) color filters on each pixel are visible.

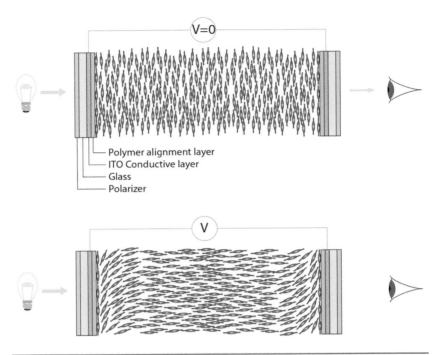

FIGURE 3.34   Schematic of a simple single LCD pixel demonstrating the response of the nematic liquid crystal on application of an electric field between the glass plates due to the applied voltage, V.

optimal properties for the intended application. LCDs can be reflective or transmissive, and there are several ways in which the nematic material can be aligned and reoriented using an electric field to achieve an "on" and an "off" state (i.e., a dark or a light pixel).

The simplest mode of LCD device uses the geometry illustrated in Figure 3.34. This mode of operation takes advantage of the change in birefringence when the material responds to an electric field. The liquid crystal molecules are initially constrained to a planar alignment by the use of surface coatings on ITO (indium tin oxide, the conductive coating) glass. When a voltage is applied across the glass plates, the resulting electric field perpendicular to the molecular orientation induces a Fréedericksz transition, and molecules with a positive dielectric anisotropy will reorient to align with the E-field.

Since the LCD was first developed, there have been many variations on the simple display mode shown. Different combinations of molecular orientations, surface treatments, and electric fields have yielded new, faster displays with higher contrast ratios. The liquid crystal material itself has also been highly optimized by the synthesis of a host of different liquid crystal molecules and the preparation of finely tuned liquid crystal mixtures. In this book, we do not discuss the many iterations of LCD improvement. Instead, we discuss one additional mode for the LCD, the *twisted nematic* display, one of the most commercially successful and long-lasting modes of operation.

3.9.2   THE TWISTED NEMATIC DISPLAY

The twisted nematic display has a design very similar to that of the simple nematic display, except for an important difference; the liquid crystal inside the pixel is not uniformly aligned in a specific direction. In fact, there is a 90° twist between the upper and the lower plates (Figure 3.35). As polarized light with an orientation matching the molecular orientation of the liquid crystal just below the upper plate passes through the liquid crystal, its polarization is effectively rotated as the molecules rotate. The twisted liquid crystal acts as a waveguide, rotating the polarization of the light as it passes through.

This 90° twisted alignment produces a bright state with no field applied. Incident polarized light is guided through the liquid crystal film. On reaching the back side of the film, the light is either reflected after passing through a second polarizer aligned with the molecular orientation on the back side or can be transmitted through that polarizer. With the application of an electric field across the conducting glass plates just as in the birefringence mode device (Figure 3.34), the molecules reorient parallel to the E-field between

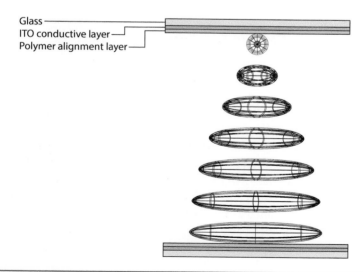

FIGURE 3.35   The geometry of a twisted nematic liquid crystal device. Alignment layers are designed such that the molecules adjacent to one glass plate are twisted 90° to those at the other glass plate.

the conducting plates (except for those close to the alignment layers), reducing the device birefringence to almost zero. Thus, a dark state is generated. Twisted nematic displays can be constructed to act in reflective or transmissive modes. They have good contrast, but a poor viewing angle, and are best viewed from straight on. Early digital watches and calculator displays in the 1970s employed this configuration, and the design is still used today in many simple displays. Other display modes, such as the *super twisted nematic*, the *in-plane device*, and the *vertical alignment display*, use different combinations of liquid crystal orientation and electric field geometry, but the principle behind all of these displays remains the same: the birefringence of a liquid crystal film can be varied by reorienting the liquid crystal molecules in an electric field.

3.9.3 SPATIAL LIGHT MODULATORS

A spatial light modulator (SLM) is used to modify a beam of light spatially as it passes through the device. When linearly polarized light is incident on a liquid crystal film, the film can act as a quarter-wave plate (retarding the light by a quarter wavelength, $\pi/2$), half-wave plate (retarding the light by a half wavelength, π), or more generally produce elliptically polarized light for an arbitrary retardation, depending on the birefringence of the material and the thickness of the film. The SLM is composed of pixels, similar to an LCD, and is used in optical applications to create a light distribution, spatially controllable in amplitude and phase.

3.9.4 LIQUID CRYSTAL TEMPERATURE SENSORS

The optical properties of the cholesteric phase provide a useful mechanism for a temperature sensor. These devices may be found in pet stores (e.g., fish tank thermometers—Figure 3.36) or the hardware store (e.g., home temperature monitors) and are extremely cheap and relatively disposable. In a cholesteric liquid crystal thermometer, a thin layer of cholesteric liquid crystal is placed on a black substrate and covered with a transparent plastic film. The liquid crystal molecules are oriented using surface coatings to have a planar alignment (as shown in the optical microscope image in Figure 3.12). The cholesteric material in such a device has a periodic variation in refractive index. In this geometry, unpolarized white light incident on the chiral structure will be reflected back as circularly polarized light. The reflection is strongly wavelength dependent, and at normal incidence the reflected wavelength is of maximum intensity when:

$$\lambda = nP \tag{3.29}$$

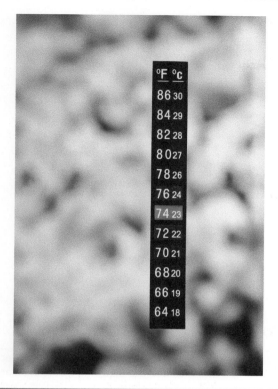

FIGURE 3.36 A fish tank thermometer uses a reflective cholesteric film as the sensor. As the water temperature changes, the cholesteric helix varies in pitch, causing a change in the reflected wavelength of the material.

where P is the pitch of the cholesteric material, and n is the average refractive index $(n_e - n_o)/2$. The reflected bandwidth is therefore given by:

$$\Delta\lambda = P(n_e - n_o) = P\Delta n \tag{3.30}$$

where Δn is the birefringence of the material. A schematic of the reflection spectrum for this cholesteric device is shown in Figure 3.37. Outside the reflection band, neither handedness of polarization will be reflected from the structure.

Figure 3.37 also shows an idealized view of a cross section of the temperature-sensitive film. A variety of different temperature-sensitive products, such as paints, coatings, and films, make use of selective cholesteric reflection for temperature measurement.

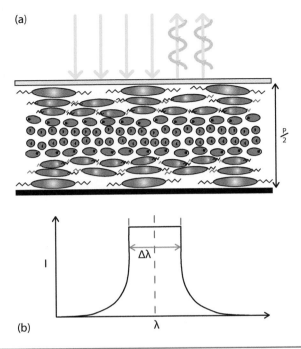

FIGURE 3.37   (a) Diagram showing the simplified structure of a reflective temperature sensor. Incident light with a handedness matching that of the material is selectively reflected from the liquid crystal structure. (b) The reflection spectrum from the
sensor. Here we plot the intensity of reflected light, I as a function of wavelength, λ

WORKED EXAMPLE 3.2

A planar aligned cholesteric liquid crystal film, such as that found in a
temperature sensor (and in Figure 3.33) selectively reflects a band of
light centered at 520 nm. If the average refractive index of the material
is 1.6, what is the pitch of the helix? If the birefringence is 0.2, what is
the bandwidth? Deduce values of the ordinary, n_o and extraordinary, n_e
refractive indices.

The width of the cholesteric reflection band is given by:

$$\Delta\lambda = P(n_e - n_o) = P\Delta n$$

where P is the cholesteric pitch, so the center of the reflection band is
found at,

$$\lambda_0 = \bar{n}P = P(n_e + n_o)$$

(Continued)

WORKED EXAMPLE 3.2 (Continued)

Using these equations, the pitch is, $P = \frac{\lambda_0}{\bar{n}} = \frac{520}{1.6} = 325$ nm,
and the bandwidth, $\Delta\lambda = P\Delta n = 325 \times 0.2 = 65$ nm.
   The two different refractive indices are obtained by using the two
equations above for $\Delta\lambda$ and λ_0, to obtain,

$$n_e = \frac{\Delta\lambda + 2\lambda_0}{2P} = \frac{65 + 2(520)}{2(325)} = 1.7$$

And

$$n_o = \frac{2\lambda_0 - \Delta\lambda}{2P} = \frac{2(520) - 65}{2(325)} = 1.5$$

QUESTIONS

The Characteristics of Liquid Crystal Materials

1. Explain to a non-scientist what a liquid crystal is. What is the difference between a thermotropic and a lyotropic liquid crystal? Give some examples of each.
2. A liquid crystal has "short-range orientational order." Explain this concept. How does this characteristic compare with crystalline solids and isotropic liquids?
3. An order parameter for a nematic phase can be described by the second order Legendre polynomial, $S = 1/2 3\cos^2\theta - 1$. Solve this equation for an order parameter $S = 1$ and $S = 0$. What is the significance of your answers?

Anisotropy and Birefringence

4. Liquid crystal phases are anisotropic, and the **c** director is used to indicate the direction of molecular orientation. In the case of the (a) columnar discotic, (b) nematic, and (c) smectic C phases, make a sketch of the molecular arrangement and indicate the **c** director.
5. Which of the following liquid crystal samples will be birefringent when viewed on the polarizing microscope between crossed polarizers: (a) homeotropically aligned nematic phase, (b) the isotropic phase, (c) a planar-aligned smectic A phase, (d) the cholesteric phase with a focal conic texture, or (e) the isotropic phase of a discotic material?

6. A thin film of nematic liquid crystal is filled between two glass parallel plates for use as a quarter-wave plate with a 488 nm laser. If the liquid crystal has a birefringence of 0.22, calculate the thickness of the liquid crystal slab

7. When a planar-aligned liquid crystal slab with a birefringence Δn and a uniform director is placed between crossed polarizers and the slab rotated between the polarizers, the total transmitted intensity varies as $\sin^2 2\theta$, where θ is the angle between the polarization direction of the first polarizer and the direction. Starting with an incident wave with intensity I_0, show that (a) the ordinary and extraordinary waves have a phase shift:

$$\Delta\phi = \frac{2\pi}{\lambda} d\Delta n,$$

on exiting the material and (b) the intensity of light passing through the second polarizer (the analyzer) follows the relation,

$$I \propto I_0 \sin^2 2\theta$$

8. Liquid crystal phases can be distinguished by examining their defect textures. What characteristic defects would indicate the presence of the nematic phase or the cholesteric phase?

Experimental Techniques and Liquid Crystal Technologies

9. A transmission x-ray diffraction experiment carried out on an unaligned smectic liquid crystal results in a single-ring pattern collected on a charge-coupled device detector placed perpendicular to the beam direction 1.5 m away from the sample using a beam wavelength of 0.154 nm. If the radius of the ring on the detector is 3 cm, calculate the smectic layer spacing for this material. Can you explain why only a single ring is observed in this system?

10. A nematic liquid crystal with a positive dielectric anisotropy of 0.05 C^2/Nm^2 is confined in a 5-μm planar liquid crystal device. If the material is observed to undergo a Fréedericksz transition with the application of a minimum of 15-V potential difference across the device, calculate the splay elastic constant K_{11}.

11. Explain how cholesteric liquid crystals can be used as a temperature sensor (liquid crystal thermometer). How does the phase change as a function of temperature?

12.   Describe a simple mode of operation for an LCD pixel. Use a diagram in your explanation.

13.   In a liquid crystal device, what is the advantage of using an initial 90 degree twist between the glass plates (i.e. the twisted nematic) over an untwisted configuration (as shown in the diagram in Figure 3.34).

REFERENCES

1.   F. Reinitzer, Beiträge zur kenntniss des cholesterins. *Monatsch. Chem.* 9, 421 (1888).
2.   O. Lehmann, Über fliessende krystalle. *Z. Phys. Chem.* 4, 462 (1889).
3.   R.B. Meyer, L. Leibert, L. Strzelecki, and P. Keller, Ferroelectric liquid crystals. *J. Phys. (Paris) Lett.* 36, L69 (1975).
4.   N.A. Clark and S.T. Lagerwall, Submicrosecond bistable electro-optic switching in liquid crystals. *Appl. Phys. Lett.* 36, 899 (1980).
5.   A.D.L. Chandani, E. Gorecka, Y. Ouchi, H. Takezoe, and A. Fukuda, Responsible for the tristable switching in MHPOBC. *Jpn. J. Appl. Phys.* 28, L1265 (1989).
6.   H.F. Gleeson and L.S. Hirst, Resonant x-ray scattering: A tool for structure elucidation in liquid crystals. *Chem. Phys. Chem.* 7, 321–328 (2006).
7.   S.-T. Wu and D.-K. Yang, *Fundamentals of Liquid Crystal Devices*. Wiley Series in Display Technology. New York: Wiley (2006).

FURTHER READING

L.M. Blinov and V.G. Chigrinov, *Electro-optic Effects in Liquid Crystal Materials (Partially Ordered Systems)*. New York: Springer (1996).

P.J. Collings, *Liquid Crystals: Nature's Delicate Phase of Matter*, 2nd ed. Princeton, NJ: Princeton University Press (2001).

P.J. Collings and M. Hird, *Introduction to Liquid Crystals: Chemistry and Physics*. Liquid Crystals Book Series. Boca Raton, FL: CRC Press (1997).

P.G. de Gennes and J. Prost, *The Physics of Liquid Crystals*, 2nd ed. International Series of Monographs on Physics. Oxford, UK: Oxford University Press (1995).

A. Jakli and A. Saupe, *One- and Two-Dimensional Fluids: Properties of Smectic, Lamellar and Columnar Liquid Crystals*. Boca Raton, FL: Taylor & Francis Group (2006).

I.-C. Khoo, *Liquid Crystals*, 2nd ed. Wiley Series in Pure and Applied Optics. New York: Wiley Interscience (2007).

Surfactants

4.1 INTRODUCTION

A surfactant, or "surface-active agent," is a molecule that tends to localize at the interface between two immiscible fluids. This behavior gives rise to a huge diversity of interesting phases when the molecules are combined with different fluids. Surfactants can act to stabilize droplet dispersions of one fluid in another, they can form their own complex phases as a function of concentration in a solvent, and they can even form incredibly thin stable monolayers at a fluid interface.

The most common example of surfactant behavior that you will be familiar with is that of a detergent-like molecule. Detergents localize at the interface between water and oil. They are able to do this because of the unique properties of many surfactant molecules, one part being hydrophilic (water loving) and another part hydrophobic (water fearing). The water-adverse part of the molecule orients into the oil phase, whereas the water-loving hydrophilic part preferentially orients into the water. Surfactants are also known as amphiphilic, which means "loving both."

The interesting behavior of surfactant molecules has been noted for a long time, although their properties were not really understood until the beginning of the twentieth century. In a classic experiment (as reported in a letter to William Brownrigg dated November 7, 1773), the American scientist Benjamin Franklin demonstrated the spreading of surfactant on a water's surface by adding a small quantity of oil to the surface of a lake. He observed that a teaspoon quantity of the oil, when applied to the water's surface, spread rapidly to cover a very large distance right across the lake. Although he did not understand the mechanism at the time, Franklin had actually observed the surfactant molecules localizing at the water–air interface and forming a monolayer, that is, a nanoscale layer just one molecule thick. The surprising effect occurred because the oily lipid molecules Franklin used were amphiphiles—molecules with both hydrophobic and hydrophilic properties.

The observation by Franklin was repeated in an experiment by Lord Rayleigh in 1890, who made a calculation of the thickness of the oil layer. During that same period a German scientist, Agnes Pockels, also made some insightful observations on the behavior of thin surface films and constructed a rudimentary device for the measurement of surface tension and the effects of surface films on this quantity.[1] Her device laid the groundwork for the modern surface film apparatus, the Langmuir-Blodgett trough, which we describe further in this chapter along with the use of this equipment and some surface tension measurement techniques.

4.2   TYPES OF SURFACTANTS

Surfactant molecules can be diverse in structure, and there are many examples of surfactants in everyday life, from the soaps we use to wash with, to the lipids that make up our cell membranes, and even to the proteins that help to stabilize the foam on your beer.[2]

The key characteristic of any surfactant is that there should be a hydrophobic section and a hydrophilic section to the molecule or even more

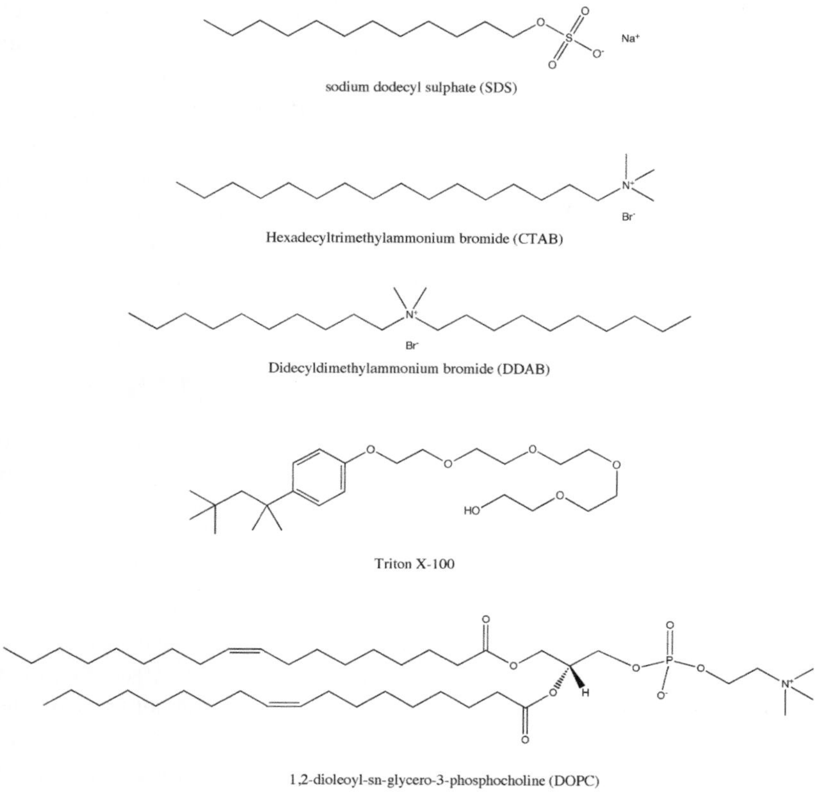

FIGURE 4.1   Some examples of common surfactant molecules, from top: anionic surfactant SDS (sodium dodecyl sulfate, also commonly known as sodium lauryl sulfate); cationic surfactants CTAB (hexadecyltrimethylammonium bromide); DDAB (didecyldimethylammonium bromide); commercial non-ionic detergent Triton X-100; and zwitterionic lipid 1,2-dioleoyl-sn-glycero-3-phosphocholine.

generally, parts of the molecule with a chemical preference for one solvent or phase over another. In many surfactants (although not proteins), this will typically be a hydrophilic "head group" with a hydrophobic hydrocarbon "tail." Some examples of surfactant molecules with this kind of structure are drawn in Figure 4.1. As you can see from the figure, each of the molecules shown has at least one long hydrophobic hydrocarbon chain.

One way to classify surfactants is by looking at the charge of the head group. An anionic surfactant has a negatively charged head group, and a cationic surfactant has a positively charged head group. There are two classifications of surfactant with a net charge of zero. If the head group of the surfactant is uncharged, as in the case of Triton X-100 (see Figure 4.1), then the molecule is a non-ionic surfactant. However, surfactants with different charges on the head group with some physical separation as in the case of the lipid DOPC (1,2-dioleoyl-sn-glycero-3-phosphocholine (Figure 4.1), but no net overall charge are called *zwitterionic*. Zwitterionic surfactants may become positively or negatively charged overall depending on the pH of the solution in which they are dissolved.

Surfactant molecules can come in a variety of different shapes and sizes, and these properties tend to affect their bulk phase behavior and their ability to form different self-assembled structures. In this chapter, we look into these effects in more detail and discuss the effects of packing constraints on phase formation.

4.3 SURFACE TENSION AND SURFACTANTS

Surface tension, the apparent force that seems to hold a droplet of liquid together, is strongly affected by the presence of a surfactant. Localizing at the interface between a fluid and the air, surfactants typically reduce the surface tension.

The surface tension γ of a liquid provides a measure of the cohesive forces between the molecules at a surface. Molecules in a liquid are attracted to each other; therefore, it takes some energy to separate one from the bulk. If we just think about a single molecule in a fluid, attractive forces will act between neighboring molecules on all sides, so the net force on the molecule is zero. On the surface of a fluid droplet or film, only molecules from within the droplet will exert these attractive forces, resulting in a net force into the bulk fluid as illustrated in Figure 4.2. This is the effect we call surface tension.

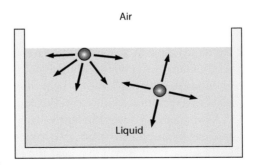

FIGURE 4.2 Surface tension results from anisotropic cohesive forces at a fluid interface. In this diagram, two molecules in a liquid are highlighted, and the cohesive forces are indicated by the arrows. The molecule in the bulk is subject to an average net force of zero due to isotropic cohesive forces from the surrounding molecules, whereas a molecule at the surface is subject to a net force into the liquid.

FIGURE 4.3   Floating a coin on water. In this experiment, the surface tension of clean water produces a force large enough to float this aluminum 1-yen coin on the surface. If we add a tiny amount of detergent (e.g., by rubbing a finger on a bar of soap, then touching the fluid surface), surface tension will be reduced, and the coin will fall.

The effects of surface tension can be easily visualized if we place a very light object on the surface of water, like the coin in Figure 4.3. You can try this experiment with any light object, such as a needle or a paper clip. Many insects, like pond skaters, take advantage of the effect to walk on water assisted by hydrophobic hairs on their feet and a very low body mass.

In the absence of any other forces, droplets of a liquid will adopt a spherical shape due to their surface tension. This shape minimizes the surface energy of the droplet by minimizing the surface area of the interface. The bulk state is the most stable environment for a molecule in a liquid phase, so whenever we create a free surface in that phase, it costs energy. We can define surface tension in terms of the energy it takes to create a surface from a bulk material.

A classic way to define surface tension in this way is to imagine forming a thin film of fluid in a rectangular wire frame in the x–y plane (Figure 4.4). The film is y wide and x long. We can increase the area of the film by expanding the frame in the x direction. The surface tension of the film can be defined as the increase in free energy per unit area because it takes energy to create

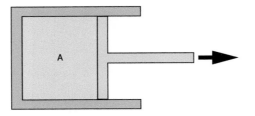

FIGURE 4.4 A simple method for spreading a fluid film in a wire frame. As the movable side is extended, the interfacial area, A, between air and the fluid increases.

this interface or area A from a bulk fluid. The surface tension is the work done to increase the area of the film by dA, or:

$$\gamma = \frac{dW}{dA} = \frac{Fdx}{ydx} \tag{4.1}$$

This means that we can also write the surface tension as:

$$\gamma = \frac{F}{y} \tag{4.2}$$

where F is the force it takes to spread a film of width y—the surface tension of water at room temperature is 0.072 N/m.

Surfactants reduce the surface tension of a liquid when they localize at the surface. Try this simple experiment at home: Take a sewing needle, a very clean beaker of water, a small piece of tissue, and some soap. Fill the beaker with water, then float a small piece of tissue (a few centimeters square) on the surface of the water. Carefully place the needle on top of the tissue, then gently prod the tissue until it starts to sink; you want the needle to remain behind on the surface of the water. The needle is light enough that the force of surface tension is enough to keep the needle floating on the surface. Next, wet the soap and generate some soapy suds, pick up some suds on your finger and gently touch them to the surface of the water well away from the needle (you may need to add suds a few times). You should notice that suddenly the needle will sink. This happens because the soap molecules form a monolayer on the surface of the water and effectively reduce the surface tension at the water–air interface. The reduced surface tension can no longer counteract the weight of the needle; therefore, it sinks.

At very low concentrations, surfactant molecules are soluble in water (or another suitable polar solvent), and the molecules move individually in the bulk fluid. Some molecules will localize at the surface of the water,

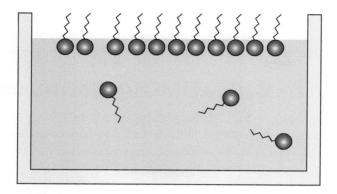

FIGURE 4.5   Surfactant molecules localize at the interface between two fluids reducing surface tension.

the surface excess (Γ), and others will be dispersed in the bulk (Figure 4.5). Molecules localized at the surface will reduce the surface tension.

The Gibbs isotherm equation relates the surface tension, surface excess, and the surfactant concentration c, where R is the ideal gas constant, and T is the temperature.

$$\frac{d\gamma}{d\ln c} = -RT\Gamma \tag{4.3}$$

From this equation, we can see that if additional surfactant is added to the system and it goes to the interface, then the surface tension will decrease.

The increase in the surface excess as a function of concentration depends on the affinity of the surfactant molecules for the surface, and this can be quantified using the Langmuir equation,

$$a_0\Gamma = \frac{K_{ad}c}{1 + K_{ad}c} \tag{4.4}$$

In this equation, a_0 is the surfactant head group area (i.e., the area taken up on the fluid surface by a single molecule), and K_{ad} is a rate constant for surfactant adsorption to the interface. By substituting the Langmuir equation into the previous equation for $\Gamma(c)$ and then integrating, we obtain an expression for the surface tension that includes the adsorption rate constant and the surfactant concentration. In this way, we can relate the addition of a particular surfactant to its effects on surface tension.

$$\gamma = -\frac{RT}{a_0}\ln\left(1 + K_{ad}c\right) + \gamma_0 \tag{4.5}$$

Here, γ_0 represents the surface tension of the solvent with no added surfactant. As the concentration of surfactant in the solvent is increased, the additional molecules preferentially locate at the surface until a maximum density is reached.

4.4   SELF-ASSEMBLY AND PHASE BEHAVIOR

The primary characteristic of surfactant molecules is their ability to locate at the interface between two different fluids. So far, we have only discussed the effects of this property at low surfactant concentrations, but at high surfactant concentrations, it is equally important.

As we learned in Chapter 1, it is unfavorable for the hydrophobic "tail" regions of surfactant molecules to be surrounded by water molecules, so they will tend to pack together. There is an energy cost associated with adding a hydrophobic molecule to water, and this problem is encountered when we try to dissolve molecules with a hydrocarbon chain in water. The non-polar section of the surfactant cannot form hydrogen bonds with the polar solvent, resulting in the formation of a shell of hydrogen-bonded water molecules around the chain. Such an arrangement of water molecules will decrease the entropy of the system and thus requires an increase in the free energy. The hydrophilic head groups of those same surfactant molecules tend to prefer the polar water environment and localize where they can form hydrogen bonds with the water molecules.

When there is a much larger number of surfactant molecules in the solvent than it would take just to form a monolayer at the surface, the surfactants in the bulk solution move randomly and adopt the most energetically favorable configuration. This self-assembly mechanism leads to the spontaneous formation of a variety of different surfactant phases driven by the hydrophobic effect (also known as "lyotropic phases" because their structure depends on concentration in the solvent). In the following sections, we examine some of the different concentration-dependent phases that can form in surfactant systems.

4.4.1   THE MICELLAR PHASE AND THE CRITICAL MICELLE CONCENTRATION

As we discovered in Section 4.3, the addition of surfactant to water reduces surface tension as surfactant molecules preferentially locate at the air–water interface in a monolayer. Above a certain concentration, we reach a point at which there are enough surfactant molecules available that they can group together to form spherical structures, or *micelles*. In these micelles, the surfactant molecules are oriented so that their hydrocarbon chains are arranged next to each other and separated from the aqueous solution by the hydrophilic head groups.

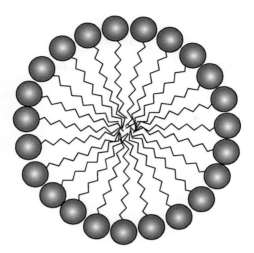

FIGURE 4.6 Idealized geometry of a micelle. The alkyl chains on the surfactant molecule are directed toward the center of the micelle; the head groups arrange to coat the outside, shielding the hydrophobic core from the surrounding water molecules.

This state is optimally achieved at low concentrations when the hydrophobic surfactant tails are arranged at the core of a sphere with their hydrophilic heads on the surface. These spherical micelles form throughout the solution, screening the hydrophobic tails from the aqueous environment. Figure 4.6 shows a schematic cross section of a spherical micelle demonstrating their geometry, although in a real micelle the chain arrangements would be much more disordered than shown here. The concentration-dependent transition from a dispersed phase to this micellar phase occurs at the *critical micelle concentration* (CMC). Above the CMC, there will still be a surface monolayer and a small number of single surfactant molecules dispersed in the bulk fluid, but most of the surfactants will be part of micelles. A micellar solution is highly dynamic, and you should not think about micelles as fixed solid particles in the solution. In thermal equilibrium, molecules may constantly leave and rejoin micelles; as in any soft system, the system fluctuates as a function of temperature.

Measuring the CMC in a surfactant solution is relatively straightforward because several macroscopic parameters, such as the surface tension, viscosity, or optical scattering properties, show an abrupt change as a function of concentration (c) at the transition point where the micelles form. A plot of surface tension as a function of ln c will reveal a clear slope change at CMC as marked in Figure 4.7 by the arrow.

Two important parameters that impact the value of the CMC in a particular surfactant solution are hydrocarbon chain length and head group charge. In general, for a given head group, longer chains will result in a lower CMC.

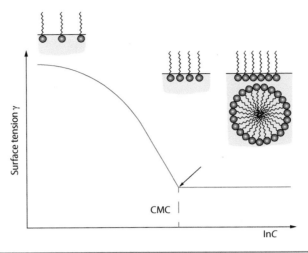

FIGURE 4.7   Surface tension in a fluid phase as a function of surfactant concentration. The arrow indicates the critical micelle concentration (CMC) at the concentration where micelles begin to form.

This effect comes from the increasing energy cost of placing longer hydrocarbon chains in water; the chains decrease local entropy by ordering the water molecules around them. For surfactant molecules to pack efficiently into a spherical arrangement, the surface area occupied by the hydrophobic heads must effectively shield the volume occupied by the tails from interactions with the surrounding water. At some optimal radius, micelles will pack into a sphere large enough that the hydrophilic heads cover the surface without significant gaps, but not so large that the inner space cannot be filled by the hydrophobic tails. The micelle is therefore a delicate balance of geometry and intermolecular forces. If the head groups cover the surface too sparsely at a given micelle radius (defined by the chain length), then the micelle will not be stable because the interior tails will be exposed to the water environment. Above chain lengths of about 16 carbons, the long chains will need to fold to fit into the interior of the micelle to preserve the hydrophilic surface layer.

In the case of anionic or cationic surfactants, the repulsive forces between head groups act against micellization, resulting in an increase in the CMC as a function of molecular ionic strength. This effect can be screened by the addition of counterions into the solution (for more detail on this effect, see Section 6.10.2).

The micellar phase is the most dilute lyotropic surfactant phase and illustrates the power of the hydrophobic effect to self-organize molecules into a structured phase. At higher concentrations, this same principle acts to induce the formation of several different surfactant phases with varying solvent fractions (Figure 4.8).

Decreasing Solvent Concentration

FIGURE 4.8 In a lyotropic system such as surfactant (white) and water (purple), different phases are stable at different volume fractions. We can relate this idea to a variety of systems, such as surfactant-water-oil mixtures or block copolymers (the polymer equivalent of a surfactant).

4.4.2 OTHER SURFACTANT PHASES

The self-assembly of surfactants is not restricted to spherical micelles, and a variety of different phases may form depending on the molecular geometry of the surfactant and its concentration in the solvent. These *lyotropic phases* may be liquid crystalline as they can exhibit short-range orientational order, so there is some significant crossover here between the fields of surfactants and the liquid crystal materials described in Chapter 3. In other texts, surfactants are sometimes discussed together under the umbrella of liquid crystal materials.

As the concentration of a micellar solution is increased, the micelles increase in concentration and start to interact with one another. At the highest concentrations of this phase, they pack closely, forming a lattice-like arrangement. Face-centered cubic (FCC) is the most efficient packing geometry for spherical particles filling 74% of the available volume, and the micelles will spontaneously arrange to form a phase with either this structure or the slightly less dense body-centered cubic (BCC) at 68%. A micellar solution can be concentrated further by evaporating away the solvent, and eventually we observe a phase transition to a new structure. The micelles merge, and the molecules rearrange into a close-packed arrangement of cylindrical micelles, allowing up to 91% of the space in the solution to be occupied by surfactant. This is also known as the hexagonal phase. Cylindrical micelles (also often known as worm-like micelles) can also occur as individual isolated tubes in some dilute solutions (Figure 4.9) instead of spherical micelles depending on the molecular geometry of the surfactant, their restriction being that the curved end caps on the micellar tubules must be able to form. Other interesting tubular shapes such as hollow lipid tubules, hollow disks, or micellar disks (lipid bicelles) can also form under controlled conditions.

The lamellar phase consists of bilayer sheets and becomes particularly important in biological systems where the surfactants present are lipids. Either side of the lamellar phase in concentration, we sometimes observe the complicated bicontinuous phases. These phases have cubic symmetry and

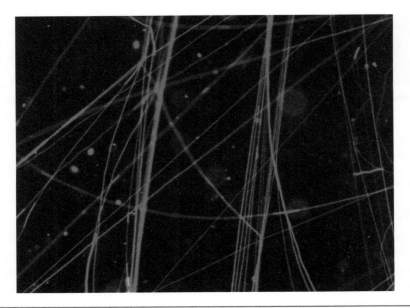

FIGURE 4.9   Bilayer tubules prepared in solution from the lipid DOPC (1, 2-dioleoyl-sn-glycero-3-phosphocholine, a biological surfactant) and imaged using fluorescence microscopy.

consist of periodically curved lamellar sheets. At higher concentrations of surfactant, we may observe formation of an inverse hexagonal phase (cylinders of water surrounded by surfactant) and even an inverse micelle phase (spherical droplets of water surrounded by surfactant). Figure 4.10 illustrates an example phase diagram for a surfactant in water.

Unlike a liquid crystal phase sequences in which different thermotropic phases occur as the temperature is increased, the usual phase sequence for a surfactant-solvent system varies with surfactant concentration as follows: micellar, cubic, hexagonal, lamellar, inverse hexagonal, inverse cubic, and inverse micellar. For any given system, not all phases will necessarily occur. Figure 4.8 demonstrates the principle behind this sequence with the solvent colored purple. A combination of volume fraction and molecular geometry drives possible phase formation. In surfactant systems as illustrated in Figure 4.10, temperature does however still play an important role. Below a certain temperature, known as the *Krafft point* the lyotropic phases do not occur—instead the surfactant is effectively insoluble in water, and crystallizes, lacking sufficient thermal energy to form mesophases sensitive to the hydrophobic effect. At sufficiently high temperatures these phases are destabilized by strong thermal fluctuations.

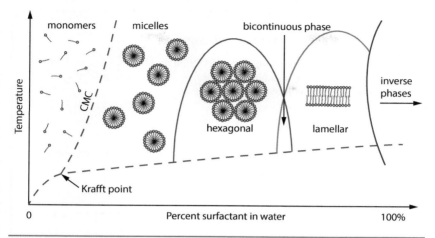

FIGURE 4.10 Example phase diagram for a surfactant/water system as a function of temperature and composition. (Inspired by Dierking, I. and Al-Zangana, S., *Nanomaterials*, 7, 305, 2017.)

4.4.3 THE PACKING PARAMETER

As we have seen in the previous section, there are several factors that determine the bulk phase of a surfactant system, including concentration in the solvent and temperature. Another key factor is the overall shape of the molecule (e.g., head group area, tail volume, and tail length). The geometric effects that overall molecular shape can have on phase behavior are summarized neatly in a quantity known as the *packing parameter*, P, where,

$$P = \frac{V}{a_o l_c} \tag{4.6}$$

In this equation, V is the average volume occupied by the surfactant molecule, a_o represents the area occupied by the molecular headgroup (hydrophilic end), and l_c is the fully extended length of the hydrophobic tail (Figure 4.11).

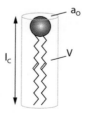

FIGURE 4.11 Geometric model for a surfactant molecule. a_o represents the area occupied by the headgroup, V is the average volume taken up by the molecules and lc is the fully extended length of the tail

WORKED EXAMPLE 4.1

Consider a micelle consisting of M surfactant molecules. Each molecule takes up a volume V and has a hydrophilic head group area of a_o. We can write the total volume of the micelle as

$$Mv = \frac{4}{3} \pi R^3,$$

where R is the micelle radius. We can also write a similar expression for the surface area of the micelle:

$$Ma_0 = 4\pi R^2.$$

By writing each of these expressions in terms of M and equating them, we obtain:

$$\frac{v}{a_0 R} = \frac{1}{3}.$$

Now, for a real molecule the maximum length of the molecule (or the critical length l_c) is equal to the radius of the micelle, so $R \leq l_c$. Therefore, we can write:

$$P = \frac{v}{a_0 l_c} \leq \frac{1}{3}.$$

P is the packing parameter,

$$\frac{v}{a_0 l_c},$$

and for micelles can be up to a value of 1/3. Above this value, it is more favorable for other structures to form if the molecules are to pack efficiently into a self-assembled structure. A similar analysis for the other surfactant phases can be performed, yielding the packing parameter values shown in Table 4.1. Carry out the same analysis for the hexagonal and lamellar phases—to verify the numbers in Table 4.1.

TABLE 4.1
Packing Parameters and Molecular Shapes for Different Surfactant Phases

Phase	Packing Parameter P	Molecular Shape
Micellar	$\leq 1/3$	Inverted cone
Cylindrical micelle (hexagonal)	≤ 0.5	Inverted cone
Lamellar	0.5 to ~1	Cylinder
Inverted hexagonal	>1	Cone
Inverted micelle	>1	Cone

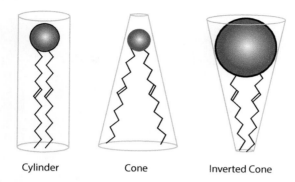

Cylinder   Cone   Inverted Cone

FIGURE 4.12   The overall shape of surfactant molecules plays an important role in molecular packing and self-assembly. If molecules are on average cone-like or cylindrical, they will pack together to produce either flat or curved geometries.

The packing parameter quantity can be used to predict the likely phases of a particular surfactant system.

The exact phase sequence for a particular surfactant in water comes about as a result of the delicate interplay between molecular geometry and the entropic effect of organizing those molecules into different phase geometries (spheres, cylinders, etc.). As for classical phase transitions, in equilibrium the surfactant/water system will tend to adopt its lowest free energy configuration at a particular temperature, pressure, etc.

The packing parameter quantifies the importance of molecular shape on phase behavior in a simple geometric argument, and although there are other factors determining phase behavior, independent of concentration, we see that the overall geometry of a surfactant molecule (for example, the volume ratio of headgroup vs. tail) has a strong influence on how molecules arrange themselves in an assembly in equilibrium (Figure 4.12).

4.5   MEMBRANE ELASTICITY AND CURVATURE

Surfactant membranes are thin sheets of molecules held together by intermolecular forces and in many cases behave like a two-dimensional (2D) fluid. The geometry of the individual molecules making up the membrane can vary greatly in terms of chain length and head group area, and these parameters are important in determining the curvature of the membrane. All surfactant monolayers can be described as having an intrinsic curvature c_0. This parameter describes the curvature of the membrane when free from external forces and results from the geometry of the lipid molecules (i.e., whether they are more cone-like, cylinder-like, etc.; Figure 4.12).

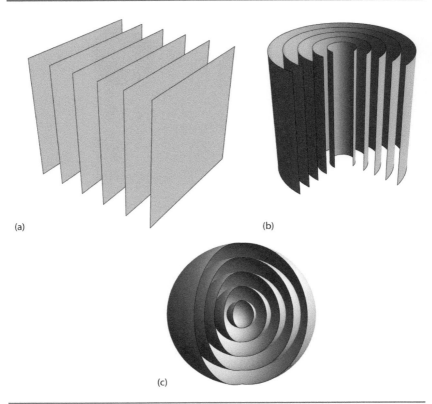

(a)

(b)

(c)

FIGURE 4.13 Diagrams demonstrating three basic multilamellar phase geometries: (a) flat, (b) cylindrical and (c) spherical. In these images, the sheets indicate the arrangement of the surfactant bilayers in a solvent medium. (Courtesy of Tim Atherton.)

If we consider how a uniform 2D membrane can be deformed into different shapes (Figure 4.13), we can describe two different in-plane axes along which the membrane can be bent, x and y, for example. The degree of bending in these two directions can be described by the principle curvatures C_1 or C_2. These curvatures are related to their associated radii of curvature (R_1 and R_2, respectively), such that:

$$C_1 = \frac{1}{R_1}, \text{ and } C_2 = \frac{1}{R_2} \tag{4.7}$$

Any curvature of a two-dimensional membrane at a point can be defined by a combination of these parameters. If C_1 and C_2 are both positive, the membrane will have a concave-like cup shape. If one curvature is negative, this will produce a saddle-like shape. Figure 4.14 gives a fun everyday example of these shapes!

FIGURE 4.14 These snacks demonstrate the geometry of two different types of membrane curvature. On the left, the bowl-shaped chip has two different positive curvatures, C_1 and C_2, whereas on the right, the saddle-shaped chip has a positive curvature C_1 and a negative curvature C_2.

In the case of a sphere, $C_1 = C_2$ everywhere on the surface; for a cylinder, we can let one of the principle curvatures equal zero (see Figure 4.13). If a membrane is deformed into a general curved shape, we can describe the elastic energy of the curved membrane as:

$$f = \int \left[\frac{\kappa}{2}(2H - c_0)^2 + \bar{\kappa}K \right] dA \qquad (4.8)$$

where H is the mean curvature, defined as:

$$H = \frac{C_1 + C_2}{2} \qquad (4.9)$$

and K is the Gaussian curvature, defined as:

$$K = C_1 C_2 \qquad (4.10)$$

c_0 is the membrane intrinsic curvature, that is, the curvature of the membrane with zero deformation (e.g., a single layer of cylindrical molecules should have an intrinsic curvature of zero, but cone-shaped molecules will pack two dimensionally into a curved sheet with no energy cost). κ is the bending modulus, and $\bar{\kappa}$ is known as the saddle-splay (or Gaussian) modulus. It is quite possible to have a curved membrane with a mean curvature of zero, i.e., in the case of saddle-like deformations, so this extra term is necessary to include the energy cost of bending a flat membrane into a saddle shape.

If the bending moduli of the membrane are known, then the energy needed to deform the membrane to a particular curved shape can be calculated using equation 4.8. The precise membrane topologies favored by a given surfactant system will depend on κ and $\bar{\kappa}$ and of course the intrinsic curvature.

4.5.1 Bicontinuous Phases

Some particularly interesting bicontinuous phases can be observed in surfactant systems close to the lamellar phase on the phase diagram. A bicontinuous phase, sometimes referred to as a microemulsion or a sponge phase, is a two-component system (e.g., surfactant and water). Both component fractions are continuous throughout the phase (i.e., not in isolated droplets or sheets). Channels of one component weave through the other, forming a continuous network with a complicated interface between the two components that represents a minimal surface. These periodic structures have a mean curvature of zero, and in these phases, the saddle-splay modulus ($\bar{\kappa}$) can become important, favoring the formation of regions of opposite principle curvatures. There are three bicontinuous structures with a cubic symmetry that can be observed in many surfactant systems, the primitive (P), diamond (D), and gyroid (G) phases. Two examples of these phases can be seen in Figure 4.15, and similar phases have also been observed in block copolymer systems, where surfactant-like polymer chains consisting of a hydrophobic block and a hydrophilic block act as surfactants and microphase separate in a similar way.

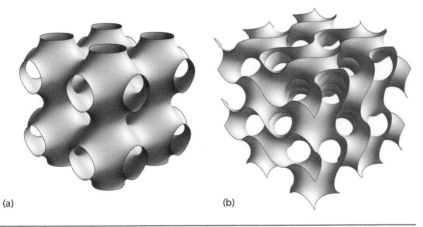

(a)   (b)

FIGURE 4.15   Two examples of bicontinuous surfactant/solvent phases: (a) the cubic bicontinuous phase, also known as the "plumber's nightmare" (Im3m), and (b) the bicontinuous gyroid phase (Pn3m). These structures are examples of the *P* (primitive) and *G* (gyroid) minimal surfaces, surfaces with cubic symmetry and zero mean curvature. These diagrams depict the geometry of the surfactant layer in a continuous solvent medium. (Courtesy of Tim Atherton.)

4.6   APPLICATIONS OF SURFACTANTS

4.6.1   DETERGENTS

Surfactants represent a class of soft materials that everybody is familiar with, although they may not know it. Whenever you take a shower or wash your hands, you take advantage of the surfactant properties of soaps to separate the oil and dirt from your skin. The most widely applied use of surfactants is as detergents. Soaps have been around for thousands of years; in fact, we know that the ancient Phoenician and Roman cultures made use of soap. The first soaps were made by heating naturally occurring fats such as coconut oil or palm oil (insoluble fatty acids) with ash (an alkali) and water. This "natural" soap dominated the industry until the early 20th century, when the first synthetic soaps were developed in response to natural oil shortages caused by World War I. By the 1940s, the synthetic sulfates had emerged as preferred detergent compounds, and they still dominate many aspects of the industry today.

Detergents are cleaning products that act by trapping and removing dirt particles from a substrate. They effectively solubilize insoluble particles so they can be rinsed away with water. The mechanism is simple to understand. Surfactants act as detergents by coating the dirty surface and trapping dirt droplets or particles inside a micelle. The hydrophobic tail of the surfactant can coat an oily dirt particle, shielding it from the water, while the hydrophilic heads point outward into the aqueous solution (Figure 4.16).

Another attribute of detergents that improves their cleaning power is the ability of surfactants to reduce the surface tension of water. For optimal

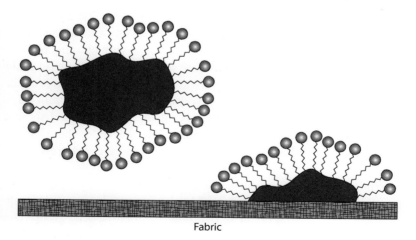

Fabric

FIGURE 4.16   The action of detergent molecules on a dirt particle. Surfactants coat the oily dirt, helping to isolate and remove it from a surface.

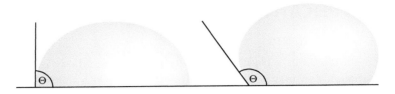

FIGURE 4.17   We can measure the hydrophobic/hydrophilic properties of a surface by looking at how water wets that surface. A drop of water placed on the surface will bead with a characteristic angle θ, the contact angle. A higher contact angle indicates a more hydrophobic surface.

cleaning, a detergent solution must wet the surface completely, and this can be a challenge for rough surfaces or the fibers of clothing. By adding detergent to water, we lower the surface tension, producing a decreased contact angle (see Figure 4.17). With a reduced surface tension, water is able to wet narrower spaces, thus delivering detergent to hidden pockets of dirt and oil trapped in the fibers of a fabric. In addition to this effect, the surface tension of water decreases with increasing temperature, so hot water wets surfaces more effectively than cold, making hot water more effective for cleaning.

4.6.2   DETERGENT FOAMS AND BUBBLES

One of the most noticeable properties of detergents is their ability to create foam, and high-foaming properties may be desirable depending on the application. For example, the ionic surfactants found in shampoos and hand dish soaps are high-foaming, but have you ever tried to add even a little hand dish soap to a dishwashing machine? This will result in large amounts of foam leaking all over the kitchen floor (unfortunately, I have tried it!). Machine detergents are in fact designed to be low-foaming agents, and in general, the presence of a high foam is not particularly correlated with cleaning power.

Foams are formed when air pockets are stabilized in a fluid phase by the presence of a surfactant, such as soap (Figure 4.18). Shampoo foam is a mixture of water and surfactant molecules (very often the anionic sodium lauryl sulfate; see Figure 4.1). When shampoo is mixed with water and sheared (lathered), air bubbles are readily incorporated into the liquid. These bubbles are then stabilized by the action of the surfactant molecules at the air–water interface.

In the same way that micelles tend to form when the hydrophobic surfactant tails find themselves dispersed in the unfavorable water environment, the soap molecules orient so that their hydrophilic head groups face each other on either side of the water film with their tails oriented away from the water in a monolayer. We will discuss some additional physics related to foams in Chapter 6—Colloids (Section 6.9)

FIGURE 4.18 Soap combined with water will form a variety of lyotropic phases. Dilute soap in water uses the micellar phase to clean, but if the soap is too dilute, it will not be effective. The concentrated soapy goop in the bottom of your soap dish is probably hexagonal or lamellar phase.

WORKED EXAMPLE 4.2: BUBBLE STABILITY

Soap bubbles in air or water are stabilized by competing forces, these forces derive from the external air pressure, internal air pressure, and surface tension. We can represent these effects by considering the total energy, dE it takes to expand a bubble by radius dR.

$$dE = P_i dV - P_e dV - 2\gamma dA.$$

The first two terms represent the work to expand the bubble by a small volume, dV, balancing internal (p_i) and external (p_e) pressures. The third term represents the surface tension energy cost to expand the membrane (see Section 3.3). We need a factor of 2 because the bubble membrane has two surfactant surfaces (enclosing a thin water shell).

Thus, for an expanding spherical bubble, $dV = 4\pi R^2 dR$, and $dA = 2(8\pi R dR)$, so we can rewrite our equation as,

$$dE = P_i 4\pi R^2 dR - P_e 4\pi R^2 dR - 2\gamma 8\pi R dR$$

or

$$\frac{dE}{dR} = 4\pi R^2 \left(P_i - P_e - \frac{4\gamma}{R} \right),$$

(Continued)

WORKED EXAMPLE 4.2: BUBBLE STABILITY (*Continued*)

at equilibrium, $\dfrac{dE}{dR} = 0$ so for a stable bubble,

$$P_i - P_e = \Delta P = \frac{4\gamma}{R}.$$

This result is the Laplace-Young equation, and it tells us something interesting about bubbles. We can expect a soap bubble with a smaller radius to have a higher pressure difference than a larger bubble. This result has important implications for coarsening (large bubbles can grow at the expense of smaller ones—see Ostwald ripening, Chapter 6, lung surfactant, Section 4.6.6), and for the surface tension measurements described in Section 4.7.2.

4.6.3 EMULSIFIERS AND EMULSIONS

One important role of surfactants in household products and foodstuffs is as an emulsifier. Emulsions are all around us in the kitchen, forming part of a wide variety of different sauces and creamy foods. When you whip air or oil into a liquid, how can the result be a stable phase? What keeps hollandaise sauce or gravy from separating into its fat and water components? The answer is an emulsifier. Emulsifiers can take a variety of forms but are essentially just surfactant molecules that localize at the interface between the aqueous and oily components of the mixture.

An emulsion is a liquid in which droplets of a second liquid are suspended (similar to the liquid–liquid colloidal suspensions, introduced in Chapter 6). The two liquids are usually immiscible (i.e., they do not readily mix like oil and water). In such a system, you should expect that if you shake the two liquids together you will form a system composed of droplets of one liquid in the other, but without some mechanism for stabilization, these liquids will eventually phase separate.

Think about the different salad dressings you see in the supermarket. Salad dressing is a mixture of oil and vinegar, and when shaken these two liquid components (oil and aqueous) mix and form a uniform cloudy suspension. But over time, the two fractions separate out again. Figure 4.19 demonstrates this effect for a simple vinaigrette salad dressing. An emulsifier will act to stabilize the suspended liquid droplets in a two-component mixture and can prevent the oil from separating out from the vinegar in your salad dressing. Surfactant molecules (which preferentially locate at the oil–water interface) are used as an emulsifier to keep the oil droplets dispersed (Figure 4.20), and therefore some salad dressings do not need to be shaken.

FIGURE 4.19   The oil- and water-based fractions of a vinaigrette salad dressing phase separate over time unless an emulsifier is used to stabilize the suspension. Take a look at the salad dressings the next time you visit the supermarket; it is easy to see which ones contain emulsifiers.

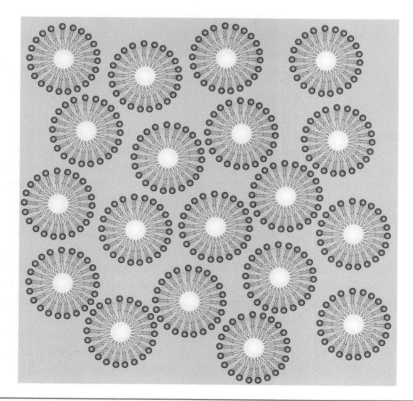

FIGURE 4.20   Surfactant molecules stabilize a dispersion of oil droplets (yellow) in water (blue).

FIGURE 4.21 Mayonnaise is an everyday example of an emulsion. Creamy foods and sauces like mayonnaise are surfactant-stabilized emulsions of oil droplets in an aqueous phase.

Egg yolk is a good example of a natural emulsifier and explains the stability of mayonnaise. Mayonnaise (Figure 4.21) consists of oil, vinegar, lemon juice, and egg yolks; to prepare the dish, the components are whisked together. Egg yolk is an important ingredient in mayonnaise because it contains a mixture of biological surfactants called lipids (see Chapter 7 for more information on lipids). These lipids (collectively known as lecithin in egg yolk) are surfactants that coat the droplets of oil that form when the mixture is whisked together, stabilizing the oil-and-vinegar emulsion. Lipid-like emulsifiers are used throughout the food industry to stabilize products consisting of oil-based and water-based ingredients (you will notice lecithin on many food labels). Their action also influences food texture, handling properties, and shelf life.

To quantify the emulsifying power of a molecule you could use the emulsifying activity index (EAI). If I add a certain volume of oil V (the dispersed phase) to water (the continuous phase) with an emulsifier, the radius of the oil droplets formed R can be related to the increase in interfacial area by the simple expression:

$$A = \frac{3V}{R} \tag{4.11}$$

The *EAI* is defined as the interfacial area created per unit mass of emulsifier added *m* as:

$$EAI = \frac{3V}{Rm} \tag{4.12}$$

The dispersed droplet size can be characterized by several methods, including light scattering (see Section 6.10.1).

4.6.4   COMMERCIAL PAINTS AND INKS

Another everyday application of surfactants is in the suspension of pigments and polymers in commercial paints and inks. Water-based paints (or emulsion paints) are composed of three components: pigment granules, a polymer, and a solvent (water). The polymer must be dispersed in the water, but when dry achieve a stable coating that cannot be washed off. To achieve this, paint is designed as an emulsion in which tiny droplets of polymer are stabilized in the solvent using a surfactant. The surfactant molecules coat the surface of the polymer drops and stabilize the material. Dye pigments are dispersed throughout and may also be stabilized using a surfactant. After application, the water solvent slowly evaporates, and the droplets eventually merge together to form a durable thin film. The surfactant allows the hydrophobic polymer base of the paint to be spread using a non-toxic water-based medium. Ink jet printer inks are based on the same dispersal mechanism. Solid pigment particles such as carbon black are dispersed in water by coating the particles with a surfactant; this allows efficient handling and deposition of pigment on the page with none of the unpleasant odors produced by other solvents.

4.6.5   SURFACTANTS AND GEL ELECTROPHORESIS

In the biological sciences, gel electrophoresis is an important technique for protein purification and identification. The goal of the technique is to determine and quantify different protein fractions present in a solution and is therefore used throughout the biochemical sciences as an important step in protein purification.

To carry out the technique, protein molecules are denatured and move through a polyacrylamide gel by electrophoresis to be sorted by molecular mass. Proteins with the highest molecular mass move through the gel more slowly under the influence of the applied electric field, but because different proteins have a variety of different shapes and charge distributions, the experiment must be carefully designed so that these factors do not affect the rate of displacement, and that mass is really the only parameter influencing migration rate.

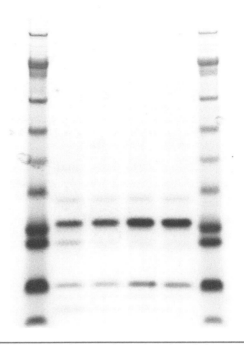

FIGURE 4.22 An example of SDS-polyacrylamide gel electrophoresis. Coomassie blue stain is used to visualize proteins as they migrate through the gel. The two vertical side bands (lanes 1 and 6) contain "ladder" proteins with known molecular masses that act as a calibration sample. The center lanes contain 6xHis-tagged murine tumor necrosis factor alpha (~19 kDa). (Courtesy of Meng-Lin Tsao.)

The typical way to control for molecular size and charge in these experiments is to carry out SDS-PAGE (sodium dodecyl sulfate [Figure 4.1] polyacrylamide gel electrophoresis; Figure 4.22). SDS is an anionic surfactant and, interestingly, the same molecule used as the primary ingredient in most commercial shampoos. Its primary function in SDS-PAGE is to denature the protein by unfolding the complex molecular structure into a long amino acid chain. SDS also coats the chain uniformly with negatively charged molecules, thus removing the effects of charge on the migration of the protein chain.

4.6.6 SURFACTANTS IN THE LUNGS

Surfactants play an important role in the lungs because they provide a mechanism to prevent droplets of water from blocking the narrow airways of the pulmonary system. By lowering the surface tension of fluids in the

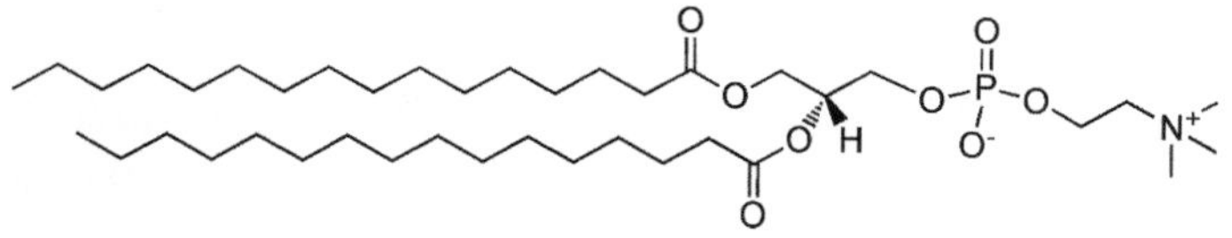

FIGURE 4.23 1,2-dipalmitoyl-*sn*-glycero-3-phosphocholine, a saturated phospholipid and the primary component of lung surfactant.

lungs, they improve wetting on the lung surface. Lung cells are coated with a water film, on the surface of which is a surfactant layer. The composition of this layer contains a large percentage of the lipid DPPC (1,2-dipalmitoyl-*sn*-glycero-3-phosphocholine; Figure 4.23), although it also contains other lipids, proteins, and cholesterol.[4]

Surfactant plays an additional important role in preventing lung collapse. The lung is filled with tiny bubble-like compartments called Alveoli. When we breathe in, these alveoli fill with air and increase in size, when we breath out, they shrink. Like a soap bubble, the alveoli experience pressure from both inside and out, and the pressure difference across their membrane can be described by the Young-Laplace equation,

$$\Delta P = \frac{4\gamma}{R} \tag{4.13}$$

where R is the radius of the spherical bubble, and γ is the surface tension of the membrane.

If we imagine the lung as a system of connected bubbles (a very simplified model), from the Young-Laplace equation it follows that small bubbles will tend to collapse and larger ones grow. Smaller bubbles have a higher internal pressure than larger ones, producing a net flow of the air inside into the larger bubbles. Such a process could be disastrous in the lungs, but fortunately surfactant plays a clever role in counteracting this effect.

The surface tension γ is related to the density of surfactant on the lung surface. When we breathe in and the alveoli expand, surfactant molecules spread out on the membrane surface, increasing surface tension. When we breath out, they shrink and the surfactant density increases, reducing surface tension. This mechanism counteracts the size-dependent pressure changes and acts to reduce the risk of lung collapse.

A common problem for premature babies is their lack of pulmonary surfactant. The pulmonary surfactant is formed late in pregnancy, and a premature baby born without the surfactant can experience respiratory distress syndrome, whereby the alveoli can collapse—a potentially fatal condition. Treatment of newborn premature babies with a lung surfactant therapy was a major advance in neonatal medicine.

4.7 EXPERIMENTAL METHODS

Many of the experimental techniques used to characterize surfactant systems are common to other areas of soft matter science and therefore are featured in other chapters of this book. X-ray diffraction, light scattering, and differential scanning calorimetry are important tools in phase identification and characterization (see Chapter 3 on liquid crystals and Chapter 6 on colloids). Atomic force microscopy can be used to investigate the topology and other surface properties of single surfactant bilayers or monolayers deposited on a surface (see Chapter 7, Section 7.5.1 on biomembranes). Spectroscopic tools such as Fourier transform infrared (FTIR), Raman scattering, and nuclear magnetic resonance (NMR) are also used in analysis of the chemical structures and dynamics of molecules; these techniques are described in Chapter 5 on polymers.

Because of the importance of surfactants in controlling and modifying the surface tension of fluids, it is appropriate here to describe methods for the measurement of surface tension and surfactant film characterization using the Langmuir trough apparatus.

4.7.1 THE LANGMUIR TROUGH

The Langmuir-Blodgett trough (LB trough) was developed by Nobel Prize-winning scientist Irving Langmuir and his collaborator Katherine Blodgett at the General Electric Company in the 1920s. This piece of apparatus is used to investigate the properties of surfactant films deposited on the surface of a liquid (typically water, but other solvents can be used). By measuring surface tension as a film is carefully compressed, information on surfactant–surfactant interactions and the phase behavior of the system can be obtained.

The first step in this experiment is to deposit a monolayer of surfactant on the liquid surface. Initially, if a very small amount of surfactant is added to the surface, there will not be enough to form a continuous monolayer. However, by carefully compressing the surfactant film using the barrier mechanism shown in Figure 4.24, the molecules become more crowded, and changes in the surface tension can be detected.

The LB trough is often used to measure surface pressure/area isotherms for a particular surfactant film. To carry out this measurement, the thin surfactant film is compressed by moving a barrier across the fluid surface at a constant rate while monitoring the surface tension. At a constant temperature the surface pressure, Π, is measured as a function of the interfacial area available to each molecule as:

$$\Pi = \gamma_0 - \gamma \tag{4.14}$$

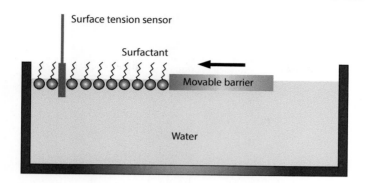

FIGURE 4.24 Schematic diagram of a Langmuir-Blodgett trough. Surfactants on the water surface can be compressed by a moveable barrier while surface tension is measured.

where γ_0 is the surface tension of the underlying solvent, and γ is the surface tension measured with the film in place. The resulting plot of pressure versus film surface area can reveal information on the phase behavior of the film, and different isotherms plotted on the same graph will indicate temperature dependence.

If a small quantity of surfactant is deposited onto a fluid surface (starting with a large area), the molecules on the surface will be well dispersed and far apart from each other; this is known as the "gas" phase. The hydrophobic tails may lie on the surface of the water, and the molecules are on average very far apart. This phase is analogous to the ideal gas we reviewed in Chapter 1 because the individual molecules are presumed not to interact with each other. As the surfactant layer is gradually compressed by reducing the film area, surface pressure is measured, and we observe different regimes in film behavior on the pressure/area isotherm. These different phases, analogous to liquid and solid bulk phases, can reveal information about surfactant and fluid interactions on the nanoscale.

Another use of the LB trough is for the carefully controlled deposition of monolayers and multilamellar films onto solid substrates, such as glass or silicon wafers. These films are known as *Langmuir-Blodgett films*. The LB trough provides a uniform monolayer on the surface of a liquid. By dipping a substrate slowly into the trough then slowly retracting, the monolayer in the trough can be transferred to the substrate. Subsequent repetitions of the dipping process will result in a multilamellar deposition.

4.7.2 MEASURING SURFACE TENSION

The Langmuir trough apparatus typically incorporates a mechanism for measuring the surface tension of the fluid interface (Figure 3.20). There are several methods used to achieve this measurement, and we review them briefly.

TABLE 4.2
Surface Tension of Some Common Liquids at 25°C

Liquid	Surface Tension (γ) at 25°C (10^{-3} N/m)
Water	71.99
Methanol	22.07
Ethanol	21.97
Mercury	485.48
Toluene	27.93
Benzene	28 22
Acetone	23.46

Each method involves exerting a force on the surface of the liquid in question and measuring the response. Surface tensions of some common liquids at 25°C are shown in Table 4.2.

One of the most popular and simple methods for making a surface tension measurement is the Wilhelmy plate or rod method. Here, either a flat plate (held vertically) or a simple rod is dipped into the fluid and then retracted slowly. The plate is designed to be very hydrophilic, so the water in the trough will wet the plate and be pulled upward as it is retracted. In this method, calculation of the surface tension is easy as the method is analogous to our previous definition of surface tension by pulling out a film from the bulk (Section 4.3).

The Wilhelmy equation relates surface tension to the contact angle θ, so:

$$\cos\theta = \frac{F}{l\gamma} \tag{4.15}$$

where l is the wetted circumference of the plate (or rod) (the length of the contact line), and F is the force exerted by the fluid on the plate. Alternatively, this method can be used to measure contact angle on a surface if the surface tension is known.

If we assume that the plate is perfectly wetted with a contact angle of 0°, then the surface tension γ is given by the force applied per unit length:

$$\gamma = \frac{F}{L} \tag{4.16}$$

This value for surface tension can then be used to calculate surface pressure.

Another immersion method that takes advantage of the immersion and retraction principle is the De Nouy ring method. In this case, instead of a plate or rod, a horizontal ring is lowered onto the fluid surface until it is slightly immersed. The ring is then slowly lifted out of the liquid and the forces on the ring measured.

An alternative method for measuring surface tension is to study the shape of a drop of the fluid, either a drop placed on a surface (a sessile drop) or a drop hanging from a pipette tip (pendant drop). In either case, a motion-less droplet of the fluid is imaged. The only forces acting on the drop are gravity and the surface tension force. The result of these forces will deter-mine the drop shape according to the Young-Laplace equation:

$$\Delta P = \gamma \left(\frac{1}{R_1} + \frac{1}{R_2} \right) \tag{4.17}$$

where ΔP is the difference in pressure between inside and outside the drop, and R_1 and R_2 are the two radii of curvature of the membrane at a point. This equation should be true at any point on the drop and using some careful non-trivial calculations of drop shape, the surface tension can be calculated.

QUESTIONS

Physical and Chemical Properties of Surfactants

1. Explain why saturated hydrocarbon chains are hydrophobic.
2. Surfactants can be anionic, cationic, or zwitterionic. Can you give examples of each?
3. How do you think a charged head group would affect (a) surfactant monolayer density and (b) surface tension?
4. Kids like to blow bubbles using a plastic or wire loop. Why is it easy to blow bubbles using soapy water, but impossible using clean water without soap?

The Hydrophobic Effect

5. Sketch a graph showing surface tension as a function of surfac-tant concentration in water. Define critical micelle concentration. why does surface tension remain constant after the CMC?
6. Placing a non-polar molecule (that can't form hydrogen bonds) into water will decrease the system's entropy. Can you explain why?
7. Explain the difference between lyotropic and thermotropic phases.

The Importance of Molecular Shape on Lyotropic Phase Behavior and Membrane Curvature

8. Consider the membrane bending energy equation:

$$f = \frac{\kappa}{2} \left(2H + c_0 \right)^2 + \bar{\kappa}_G K,$$

where H is the mean curvature, c_0 is the intrinsic curvature, and K is the Gaussian curvature. κ and $\overline{\kappa}_G$ are the bending and saddle-splay moduli, respectively.

Show that by starting with a large uniform membrane sheet (i.e., we are using a continuum model and ignoring molecular details), the energy to form a unilamellar vesicle from the sheet is not dependent on radius. Now, consider a membrane unilamellar tubule in the same way—is radius important in this case?

9. Show that the packing parameter for cylindrical micelles cannot be more than 0.5.

10. What is the difference between mean curvature and Gaussian curvature for a surfactant membrane? Describe a shape or phase for which the mean curvature is equal to zero, but the Gaussian curvature is non-zero.

REFERENCES

1. A. Pockels, Surface tension. *Nature* 43, 437–439 (1891).
2. D.J. Durian and S.R. Raghavan, Making a frothy shampoo or beer. *Physics Today* 62–63 (2010).
3. I. Dierking, and S. Al-Zangana, Lyotropic liquid crystal phases from anisotropic nanomaterials. *Nanomaterials* 7, 305 (2017).
4. R. Veldhuizen, K. Nag, S. Orgeig, and F. Possmayer, The role of lipids in pulmonary surfactant. *Biochim. Biophys. Acta* 1408, 90–108 (1998).

FURTHER READING

H.-J. Butt, K. Graf, and M. Kappl, *Physics and Chemistry of Interfaces*. Weinheim, Germany: Wiley-VCH Verlag (2004).

S. Damodaran, K. Parkin, and O.R. Fennema, *Fennema's Food Chemistry*, 4th ed. Boca Raton, FL: CRC Press (2008).

I.W. Hamley, *Introduction to Soft Matter: Polymers, Colloids, Amphiphiles and Liquid Crystals*. Chichester, UK: John Wiley & Sons (2000).

J.N. Israelachvili, *Intermolecular and Surface Forces, Second Edition: With Applications to Colloidal and Biological Systems*. New York: Academic Press (1992).

R.A.L. Jones, *Soft Condensed Matter*. Oxford Master Series in Condensed Matter Physics, Vol. 6. Oxford, UK: Oxford University Press (2002).

R.G. Laughlin, *The Aqueous Phase Behavior of Surfactants*. London, UK: Academic Press (1996).

M.J. Rosen, *Surfactants: Fundamentals and Applications in the Petroleum Industry. Surfactants and Interfacial Phenomena*. New York: Wiley (2000).

S. Safran, *Statistical Thermodynamics of Surfaces, Interfaces, and Membranes*. Boulder, CO: Westview Press (2003).

T.A. Witten and P.A. Pincus, *Structured Fluids: Polymers, Colloids, Surfactants*. Oxford, UK: Oxford University Press (2004).

Polymers

5.1 INTRODUCTION

Of all of the soft materials surrounding us in our everyday lives, polymers are by far the most common and familiar. Polymeric materials may not instantly occur to you as "soft," but hard plastics are just one of the many forms polymers can take as they are used to produce the manufactured goods that dominate our modern world. In their amorphous solid state, bulk polymers are used to create a variety of hard plastic objects, but these materials may also be designed to produce the elastic rubbers used in tires and waterproof coatings, or they can be dissolved in a solvent to create gels or viscoelastic solutions for food products and cosmetics. As with most materials, the chemical structure of a polymer can be finely tuned to produce the desired bulk properties, and throughout this chapter, we take a look at a variety of polymer structures and their uses. We start by looking at the properties of dilute polymers in solution, then bulk polymers in the glassy and melt states. Common experimental techniques used in polymer science for materials characterization are also described. Polymer molecules are very long and can behave in interesting ways because they can adopt a large number of chain configurations and at high enough concentrations become entangled with each other—they don't behave like small molecules in a simple liquid. As a result, we need to use some ideas from statistical mechanics to describe average polymer structures and their behavior in solution. For example, in a particular polymer solution it is highly unlikely that any two molecules will have the exact same conformation at a moment in time. Instead, to describe the state of the polymer molecules we think about how to describe their most likely conformation.

5.2 EARLY POLYMERS

It would be difficult in the twenty-first century for us to imagine a world without synthetic polymers. Take a quick look around your home or office and think about what your life would be like without any plastics. Polymers are all around us and have proved to be one of the most useful and durable materials in the modern world (Figure 5.1). Their durability is perhaps both a blessing and a curse, however, as we are now faced with increasing amounts of waste plastics contaminating the natural environment. It is amazing to think that just 100 years ago the average person lived in a world without synthetic polymers.

Although many natural materials can be considered polymers, for example, cellulose (starch), silk, and gelatin, the first artificial polymer materials were not manufactured until the late 19th century. One of the earliest manufactured polymers was celluloid. This material is well known

FIGURE 5.1   Polymers are one of the most common and versatile synthetic materials. They can be formed into almost any shape imaginable.

for its use in the motion picture and photography industries and was developed in the 1860s from nitrocellulose and camphor. At the time, celluloid was also widely used for a variety of simple household objects, such as combs, eyeglass frames, and piano keys, as a replacement material for expensive natural products such as ivory. Unfortunately, there was a significant drawback to the use of celluloid for everyday items; the material is flammable and liable to soften under heat. In fact, the intense light of the movie projector could cause film to burst into flames or melt unless precautions were taken. More modern film is printed on cellulose acetate (or "safety film"), a much less flammable alternative replacing the original celluloid by the 1950s.

Bakelite was the first truly synthetic polymer to be produced and was trademarked in 1907 by Belgian scientist Leo Hendrik Baekeland. This new material was formed from a combination of phenol and formaldehyde and as an early plastic enjoyed wide use in a variety of applications, from household objects, knobs, and radio casings to costume jewelry and telephones. The new plastic was hard wearing and chemically resistant. It could be infused with filler materials such as sawdust or fabric and easily molded to the desired shape. Nylon was another early and hugely successful polymer. Developed in the 1930s, this extremely durable and malleable polymer was particularly useful during World War II for its ability to be drawn into very strong fibers for parachute production.

5.3 POLYMER STRUCTURE

So, what exactly defines a polymer? A polymer is a long-chain molecule composed of a large number of repeated subunits called monomers. These subunits can be as simple as a CH_2 group, as is the case for polyethylene, or the subunit can be a much more complex collection of atoms. Even molecules as complex as proteins are polymers, long chains of different amino acid subunits. The definition of polymer can therefore be applied to an extremely wide variety of molecular structures, so the bulk properties of different polymers are diverse. The chemical structures for some examples are shown in Figure 5.2.

You will be familiar with many common, everyday polymers, such as polyethylene plastic grocery bags or the polystyrene used for hard plastics and packing foams. These molecules have a linear chain structure with a single repeating unit (e.g., AAAAAAAA) and a carbon-based backbone; they are *homopolymers*. Other non-carbon-based homopolymers exist; for example, silicone rubbers such as polydimethylsiloxane (PDMS) are polymers with a silicon backbone. A diverse array of complex chain structures are found for biological polymers. Starch is composed of a chain of glucose monomers (a polysaccharide) and is used by green plants to store energy. Even DNA (deoxyribonucleic acid) is a polymer, a long chain of four different monomeric base pair units connected to a sugar-phosphate backbone.

When polymers assemble into a long chain, they can vary along their length in different ways. For example, the chain may be composed of identical monomers, but the orientations of these monomers can vary if there is more than one way in which they can attach to the chain. This variation in structure is known as steric disorder or tacticity. If there is only one way for the monomers to couple together, then the polymer is isotactic.

Another way in which the polymer can vary along the chain is by composition; not all polymers are composed of a single monomer repeat unit, and in fact it is very common for a polymer chain to consist of a combination of two or more different repeat units. These materials are known as *copolymers* and can be either regular in their monomer repetitions (ABABABABABABAB) or random (ABBBAABABABABAAABABB, for example) or a *block copolymer* (AAAAAAABBBBBBB) composed of two or more chain segments of differing uniform composition (Figure 5.3).

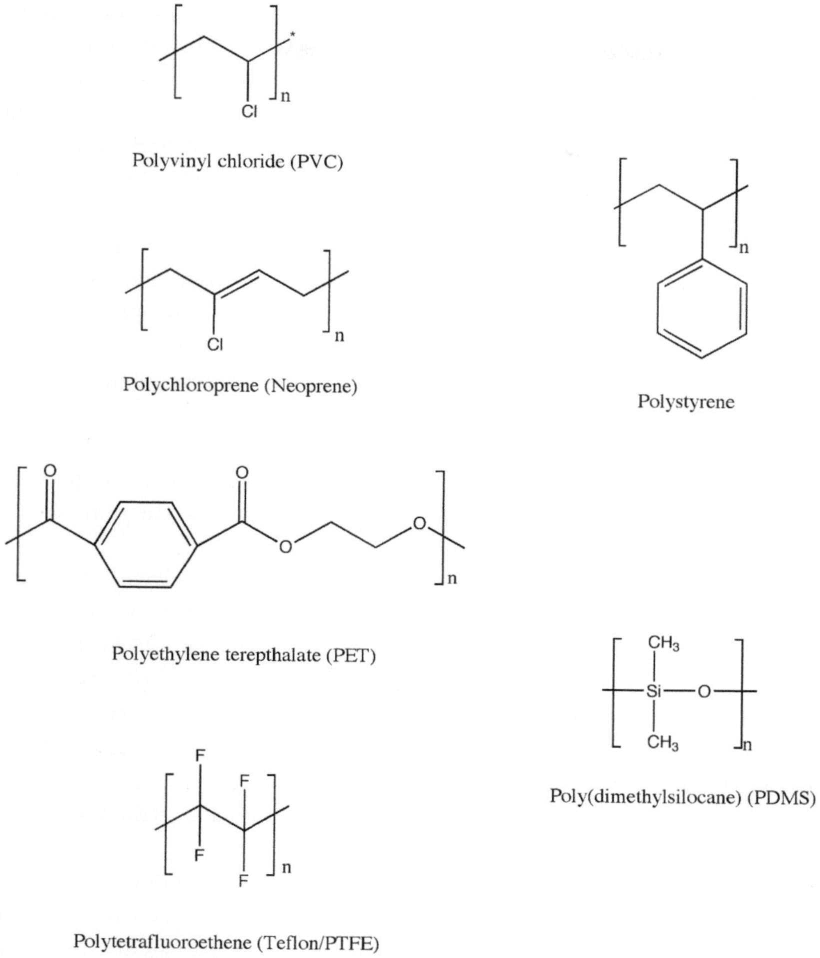

Cl
n
Polyvinyl chloride (PVC)
n
Polystyrene
Cl
n
Polychloroprene (Neoprene)
O
O
O
O
n
Polyethylene terepthalate (PET)
CH3
Si
O
CH3
n
Poly(dimethylsilocane) (PDMS)
F
F
F
F
n
Polytetrafluoroethene (Teflon/PTFE)

FIGURE 5.2 Chemical structures for some common polymers. The section of the structure shown in square brackets is repeated along the polymer chain.

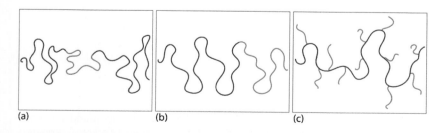

(a)
(b)
(c)

FIGURE 5.3 Cartoons demonstrating examples of different copolymer structures: (a) a triblock copolymer, (b) a diblock copolymer, and (c) a graft copolymer.

A variety of chemical synthesis methods can be used to achieve these different molecular structures, and the physical properties of a random polymer compared to the block copolymer version will vary significantly.

The interactions of polymer molecules with water or any other solvent are particularly important for the solution behavior or surface properties of the material. Solubility of the polymer chain is also critical for processing the material and therefore must be well understood. In Section 5.4, we discuss the behavior of polymers in solution and the role of the solvent. Polymers can be designed to exhibit hydrophilic or hydrophobic properties, depending on their ability to form hydrogen bonds. The hydrophobic effect we discussed in Chapter 4 in the context of surfactants applies similarly to polymer chains. If you look at the two different chemical structures shown in Figure 5.4, you will notice that the long polyethylene chain is non-polar and subsequently cannot form hydrogen bonds with water molecules, but the polar polyethylene glycol can. *Block copolymers* can even be designed to incorporate both a hydrophobic and a hydrophilic block on the same chain, resulting in surfactant-like behavior.

For example, the polyethylene–*block*–poly(ethylene glycol) molecule shown in Figure 5.5 is a block copolymer, one section of which is the hydrophobic polyethylene (PE), a simple hydrocarbon chain and our material of choice for plastic grocery bags. The other section is water-loving polyethylene glycol (PEG). When mixed with water, this block copolymer shows interesting surfactant-like behavior, and a variety of different self-assembled phases can be observed as a function of the relative block lengths and concentration in water. This behavior is very reminiscent of the surfactant phase diagram we looked at in Chapter 4.

Other variations in structure include branched polymers, star polymers, and graft copolymers. A graft copolymer is composed of a polymer backbone with a second, different polymer attached repeatedly as a side chain; an example of this molecular structure can be seen in Figure 5.5.

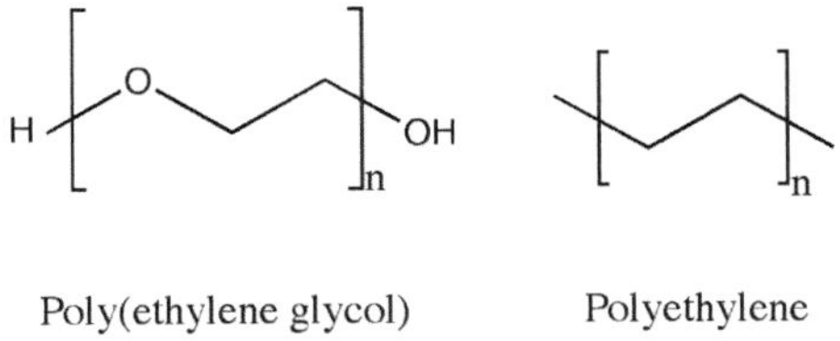

Poly(ethylene glycol)   Polyethylene

FIGURE 5.4   Examples of the molecular structures of a hydrophilic and a hydrophobic polymer. The section of the structure shown in square brackets is repeated along the polymer chain. Poly(ethylene glycol) is a hydrophilic polymer; polyethylene is hydrophobic.

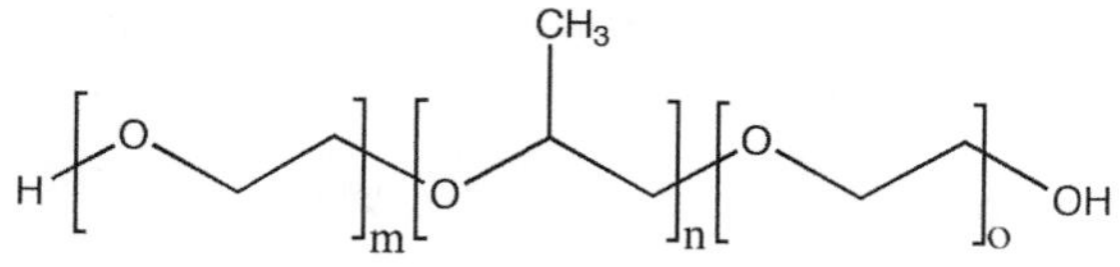

Poly(ethylene glycol)-block-poly(propylene glycol)-block-poly(ethylene glycol)

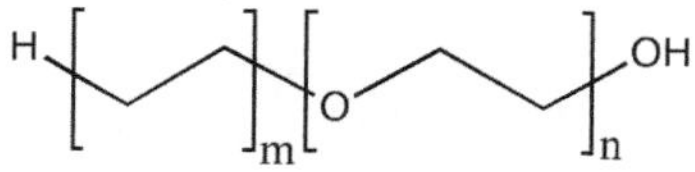

Polyethylene-*block*-poly(ethylene glycol)

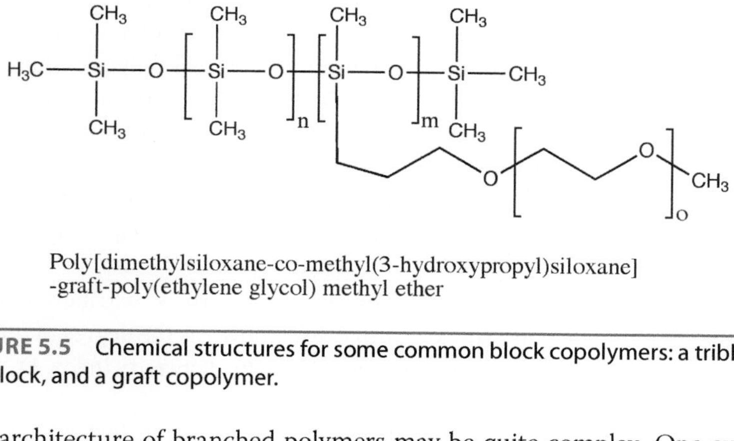

Poly[dimethylsiloxane-co-methyl(3-hydroxypropyl)siloxane]
-graft-poly(ethylene glycol) methyl ether

FIGURE 5.5   Chemical structures for some common block copolymers: a triblock, a diblock, and a graft copolymer.

The architecture of branched polymers may be quite complex. One example of a branched polymer is the biologically important molecule glycogen. Glycogen is a highly branched polysaccharide found in animal cells and used as an energy storage mechanism.

All polymers are characterized by their very high molecular mass. They are macromolecules typically composed of thousands of subunits, but between different polymer types there can be a huge range in molecular mass, from about 1000 Da* to more than 1,000,000 Da. Shorter chains of monomers with less than about 20 repeat units are referred to as *oligomers*.

---

* A Dalton (Da) is a unit of molecular mass often used for large molecules. One Dalton is equal to 1/12 of the mass of a free carbon atom. One Dalton is equal to one atomic mass unit, u, with a value of 1.66053886 (28) × 10⁻²⁷ kg.

We can write the molecular mass of an N-unit polymer as simply the sum of the mass of N subunits.

$$M = \sum_{i=1}^{N} M_i \tag{5.1}$$

where M_i is the molecular mass of a monomer. So, for a polymer composed of two different monomers, this would give a total mass of:

$$M = \sum_{i=1}^{A} M_i + \sum_{j=1}^{B} M_j \tag{5.2}$$

where A is the total number of monomer A in the chain, and B is the total number of monomers B. M_i is the molecular weight of monomer A, and M_j is the molecular weight of monomer B. A calculation using Equation (5.2) would not include any chemical end groups that vary from the main monomer chain, and for a highly branched polymer this effect may be significant. The formula also applies only to materials composed of polymer chains all of the same length (a uniform polymer). In any real polymer, there will typically be a broad distribution in molecular weights from molecule to molecule (a non-uniform polymer). This characteristic is often referred to as the *polydispersity* or *dispersity* of the material. Because of this length distribution, a specified molecular weight for a polymer is usually representative of the average molecular weight for the macromolecules that comprise that polymer system.

Most synthetic polymers are non-uniform, with a distribution in N (number of monomers in a single chain)—also known as the degree of polymerization. We can quantify this dispersity using the dispersity index, Đ.

$$Đ = \frac{\bar{M}_w}{\bar{M}_N} \tag{5.3}$$

In Equation (5.3), we look at the ratio between M_w, the weight averaged molecular mass of the polymer molecules and M_N, the number averaged mass of a polymer molecule. The number average mean is the arithmetic mean value for number of monomers in a polymer chain.

$$\bar{M}_N = \frac{\sum n_N M_N}{\sum n_N} \tag{5.4}$$

where n_N is the number of molecules consisting of N monomers and M_N is the mass of an N segment chain. This means in the polymer, there are equal numbers of molecules on either side of M_N in the mass distribution. The weight averaged mass is calculated as follows,

$$\bar{M}_w = \frac{\sum w_N M_N}{\sum w_N} = \frac{\sum n_N M_N^2}{\sum n_N M_N} \tag{5.5}$$

where w_N is the weight of a chain, N monomers long. This calculation will produce a higher average mass than that calculated for M_N. There is an equal weight of polymer on either side of M_w in the distribution. It makes sense therefore that $M_w > M_N$ as the longer chains have a greater mass.

WORKED EXAMPLE 5.1: CALCULATING POLYMER DISPERSITY

An imaginary polymer material consists of 100 macromolecules. The molecules are composed of a single monomer type with a distribution in degree of polymerization, N summarized in the table below:

Number of Macromolecules	Degree of Polymerization
8	200
22	750
45	820
21	680
4	230

To calculate the number averaged mass, we do:

$$\bar{M}_N = \frac{\sum n_N M_N}{\sum n_N} = \frac{8(200) + 22(750) + 45(820) + 21(680) + 4(230)}{100}$$

$$= \frac{70200}{100} = 702$$

Now to calculate the weight averaged mass,

(Continued)

**WORKED EXAMPLE 5.1: CALCULATING
POLYMER DISPERSITY (*Continued*)**

$$\bar{M}_w = \frac{\sum w_N M_N}{\sum w_N} = \frac{\sum n_N M_N^2}{\sum n_N M_N}$$

$$= \frac{8(200)^2 + 22(750)^2 + 45(820)^2 + 21(680)^2 + 4(230)^2}{8(200) + 22(750) + 45(820) + 21(680) + 4(230)}$$

$$= \frac{52875000}{70200} = 753.2$$

Finally, the dispersity index, Đ is given by:

$$Đ = \frac{\bar{M}_w}{\bar{M}_N} = \frac{753.2}{702} = 1.07$$

Note that as expected, the weight averaged molecular mass is higher than the number averaged molecular mass.

5.4   POLYMER SOLUTIONS

So far in this chapter, we have considered polymer chains as a rather abstract concept, but not looked at how they form real materials, interact with each other, or dissolve into solutions. In this section, we will give some thought to the behavior of polymer chains in fluids. The first question to ask is, how much space does a polymer take up in solution? Is the long chain extended? Coiled up into a tightly packed globule? Or something in between? This is actually a somewhat complicated question and in general we need to take a statistical approach. We will aim to describe the *average* behavior of polymers in solution.

If a macromolecule is placed into a fluid environment (e.g., water or some other suitable solvent), the long flexible chain can adopt a very large number of possible configurations. There will be interactions between monomers and solvent molecules and in addition, because of the long linear geometry of the polymer, there may be significant monomer–monomer interactions between sections of the same chain. In the following section, we discuss how to model the behavior of a single polymer chain, then consider the implications of solvent and monomer interactions in dilute polymer solutions.

5.4.1 THE IDEAL CHAIN

The simplest model for a polymer in solution is known as the *ideal* or *Gaussian chain*. In this model, we do not take into account any monomer–monomer or solvent–monomer interactions. A simple model for a polymer is as a *freely jointed chain*; imagine a chain of rigid rod segments, of length, a, connected together end to end, with no restriction on flexibility or angles of movement in the chain. This model implies that there should be no monomer-monomer correlations between any two sections of the chain and is a good first estimate for our polymer molecule. In solution, the ideal polymer chain can, in theory, adopt any configuration; it can be fully extended or coiled into a compact structure. In the thermal environment of the solvent, the chain will constantly undergo conformational changes at a rate related to temperature, and for a long-chain polymer, there are a huge number of possible configurations for the molecule to adopt. We can picture the polymer chain under constant bombardment by solvent molecules, so in a state of constant motion. We say that the *conformation* of the polymer takes a *random walk*. From monomer to monomer the chain takes steps in random directions, to give its exact conformation. This random conformation is similar to the path taken by a Brownian particle in solution.

In a solution containing thousands of polymer molecules, each constantly fluctuating, a huge variety of molecular shapes will be present at any given time. We can, however, try to write down the average volume taken up by a polymer molecule in a given solution at some temperature T. To do this, we first need to define the polymer's linear size, so as a start, let's consider the average end-to-end distance.

If you were to take a snapshot of all the polymers in a dilute solution at some time, the average end-to-end distance for the polymers in the system could be computed by measuring the distance between the two ends of the polymer for each molecule (Figure 5.6) and then finding the average over all molecules. If we model our polymer as a chain of N simple rod segments, the end-to-end vector for a single polymer chain is equal to:

$$R = \sum_{i=1}^{N} a_i \tag{5.6}$$

where a_i is a vector from segment i to segment $i + 1$. This vector is not particularly useful, however, to describe the size of the polymer chain, because

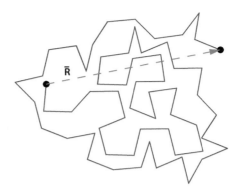

FIGURE 5.6 Definition of the end-to-end vector *R* for a polymer molecule.

the average vector over many segments will eventually tend to zero (since no particular direction is preferred—the chain distribution is isotropic), $\langle R \rangle = 0$.

Since we are using vectors to describe the path from monomer to monomer, over the entire chain of N segments, a better way to describe the end-to-end vector is to use the mean-squared end-to-end distance,

$$\langle \boldsymbol{R}^2 \rangle = \left\langle \sum_{i=1}^{N} \sum_{j=1}^{N} \boldsymbol{a}_i . \boldsymbol{a}_j \right\rangle \tag{5.7}$$

$$= Na^2 + \left\langle \sum_{i \neq j}^{N} \boldsymbol{a}_i . \boldsymbol{a}_j \right\rangle \tag{5.8}$$

Here, the first term represents the dot product of each short vector with itself and the second term averages out to zero, given an isotropic distribution of vectors (over a large number of segments). In the case of the ideal chain, we can now write,

$$\langle \boldsymbol{R}^2 \rangle = Na^2 \tag{5.9}$$

where $\boldsymbol{R}$ is the end-to-end vector for the polymer chain. This formula describes how the polymer fills space as the number of segments increases and therefore can provide a useful measure of the *average* volume taken up by the macromolecule. From this equation, we can write that the root-mean-squared end-to-end distance,

$$R_{RMS} = N^{1/2}a \tag{5.10}$$

So, the radius of the volume occupied by the polymer grows as the number of segments, $N^{1/2}$. This result can apply to the entire polymer molecule, or just to a subset of segments, to see how much space is taken up. You should notice here the parallels between the previous argument and our discussion in Chapter 2 on scaling and the fractal dimension (see Section 2.3.1). Given a large enough polymer chain, we can consider the exponent v as a scaling dimension and more generally write:

$$R_{RMS} \propto N^v \tag{5.11}$$

An exponent of more than 1/3 makes sense here for a three-dimensional (3D) polymer in solution because the polymer fills a volume space incompletely (a solid material fills space with $v = 1/3$). This ideal chain model for our polymer is modified, however, when we start to think of then behavior of a real polymer chain, for which there can be monomer–monomer and monomer–solvent interactions.

5.4.2 THE RADIUS OF GYRATION

Another measure of the space taken up by a polymer in solution is the *radius of gyration*. The radius of gyration is the effective radius of the polymer coil in solution and is defined as:

$$R_g^2 = \frac{1}{N} \sum_{i=1}^{N} \left\langle \left| r_i - r_c \right|^2 \right\rangle \tag{5.12}$$

where r_c represents the position vector of the center of mass of the polymer coil, and r_i is the position vector of the ith segment out of a total of N segments (Figure 5.7). Radius of gyration is a concept borrowed from classical mechanics of rotating bodies. The moment of inertia of a rotating continuous body is equal to the moment of inertia of a point mass placed at the object's radius of gyration. The polymer chain is of course not rotating like a solid object, but R_g is nonetheless useful as a number that describes the distribution of mass in an object, and in our case the polymer chain in space.

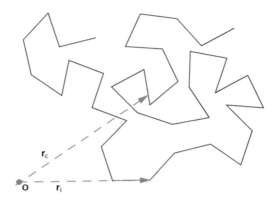

FIGURE 5.7 Parameters defining the radius of gyration R_g for a polymer molecule. r_c represents the position vector of the center of mass of the polymer coil, and r_i is the position vector of the ith segment.

WORKED EXAMPLE 5.2: HOW MUCH SPACE DOES AN IDEAL CHAIN OCCUPY IN SOLUTION?

Given a polymer of 10,000 identical segments, each spaced 0.5 nm apart, estimate the average volume occupied by the polymer chain.

Using the equation, we obtained in the section above for the root-mean-squared end-to-end distance,

$$R_{RMS} = N^{1/2} a$$

and assuming the polymer behaves as an ideal chain, this becomes a simple calculation.

$$R_{RMS} = (10,000)^{\frac{1}{2}} \times 0.5 = 50 \text{ nm}$$

From this radius, we can estimate a volume of

$$V = \frac{4}{3} \pi R_{RMS}^3 = 5.24 \times 10^5 \text{ nm}^3.$$

5.4.3 EXCLUDED VOLUME AND SOLVENT EFFECTS

There are two competing effects that predominantly modify the volume taken up by an N segment polymer in solution. The first of these is *self-avoidance*. The polymer chain can only occupy a certain point in space once as it coils around on itself; therefore, the total space available for the polymer to move is decreased. This means that there is an *excluded volume* effect that effectively increases the total space filled by the polymer chain. The second effect comes from attractions between different monomers in the chain. The van der Waals force creates an attraction between monomers as different sections of the same chain wrap around close to each other. This attraction tends to decrease the polymer's overall volume.

For a particular solvent at the *theta temperature* Θ, the excluded volume effect can be neglected, and the polymer will behave like an ideal chain with a scaling exponent of $v = 0.5$ (see Section 5.4.1). This temperature can be considered analogous to the *Boyle temperature* in a gas. At the Boyle temperature, a non-ideal gas will obey the ideal gas law, and the effects of molecular volume can be neglected.

In a *theta solvent*, the polymer will behave like an ideal chain. If a polymer is dissolved in a particular solvent at temperature Θ and the temperature of the solution is decreased so $T < Θ$, then the polymer coil will shrink. If the temperature is increased to $T > Θ$, the polymer coil will expand to fill a larger volume. The competing effects of self-avoidance and the van der Waals attraction have an impact on the density with which the polymer fills space, resulting in changes in the v exponent. The v exponent will only be equal to 0.5 in a theta solvent, but at high temperatures, we can use the Flory approximation to estimate the value of v for a self-avoiding chain,

$$v = \frac{3}{d+2} \tag{5.13}$$

where d is the dimensionality of the system, so in three dimensions:

$$R \propto N^{3/5} \tag{5.14}$$

5.4.4   INCREASING THE CONCENTRATION OF A POLYMER SOLUTION

In the previous sections, we described a polymer as a free chain, interacting only with the solvent molecules around it. However, this model is only really useful in dilute solutions. As the concentration of polymer c in a solvent is increased, the individual polymer chains increase in density until they begin to interact with each other. When a solution starts to become crowded, chain-chain interactions begin to affect how much volume the macromolecules occupy, and the isolated chain model breaks down. We can define

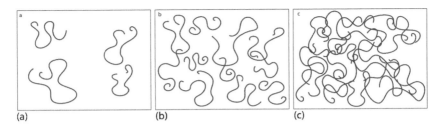

FIGURE 5.8 Representative cartoons of the three different regimes in polymer solutions with increasing concentration: (a) dilute, (b) semi-dilute, and (c) entangled polymer solutions.

three different regimes to describe how polymers behave in solution: the dilute, semi-dilute, and entangled regimes (see Figure 5.8).

Dilute: In a true dilute polymer solution, the concentration is low, and the individual polymer molecules are considered not to interact with each other. Molecules behave as self-avoiding chains and are folded up on average into a globular configuration, the size of which may be characterized by the radius of gyration R_g or the root-mean-squared end-to-end distance R_{RMS}. The total solution volume is greater than the number of molecules multiplied by R_g^3.

Semi-dilute: In the semi-dilute regime, the concentration c is approximately equal to c^*, the overlap concentration. The polymers are still on average globular; however, the volume is filled with polymers (i.e., every volume R_g^3 in the solution contains about one polymer molecule), and some chain interpenetration occurs. In practice, this regime begins at volume fractions of approximately 1%. We can estimate the semi-dilute concentration threshold in molecules per unit volume as:

$$c^* \approx \frac{1}{R_g^3} \tag{5.15}$$

Entangled: This is a concentrated solution in which the polymer chains are extended and wrapped around each other in an entangled network. The limit of this regime is a pure polymer material with a concentration c_{max}. This is a polymer melt.

The details of diffusion and scaling in these three different regimes is treated mathematically in detail in Chapter 4 of *Structured Fluids: Polymers, Colloids, Surfactants*.[1]

5.4.5   STRETCHING A POLYMER CHAIN: THE ENTROPIC CHAIN

A single polymer chain in solution will adopt a coiled conformation in the absence of any additional forces with some characteristic end-to-end distance, R. If a force is applied to the polymer, effectively opening the molecule out as indicated in Figure 5.9 and increasing the end-to-end distance by ΔR, then a restoring force will act to oppose the stretching force and restore the polymer chain to its original coiled state. At first, it is not clear exactly why this should be so; the polymer chain is being opened out into a linear configuration, but the applied force is such that the molecular bonds are not being stretched between monomers.

The explanation for this effect can be found if we consider the polymer chain using some concepts from statistical mechanics and entropy. As we described in our ideal chain model, the polymer can be described as a linear chain of N segments, each of length a. Adjacent segments can adopt any angle in relation to each other. For a given molecular end-to-end distance, R, in the case of a non-stretched polymer, there will a very large number of possible molecular conformations that will yield the same value of R. Now, think about stretching the polymer out completely into a perfect straight line. There is only one possible configurational state in which the polymer can have the end-to-end distance, $R = Na$.

Here, we can bring in the concept of entropy and the *second law of thermodynamics*—systems tend to evolve toward equilibrium, where they will be found in their most likely configuration (i.e., maximum entropy state). The equilibrium state of the polymer chain is the most likely value of R, i.e., a globular conformation. As a polymer is stretched, a decreasing number of conformational states are capable of producing the new R values. As the chain gradually adopts a less and less likely configuration (i.e., it is more and more stretched), the entropy of the chain decreases.

The entropy of an ideal chain of size R is defined by:

$$S = k_B \ln\Omega(R) \tag{5.16}$$

FIGURE 5.9   The force it takes to stretch out a polymer derives from the associated decrease in chain entropy. F here represents an applied force stretching out the polymer chain.

where Ω is the *multiplicity*, essentially the number of configuration states available to the polymer for a particular value of R (see Appendix D). By considering the polymer as an ideal chain, taking a random walk, we can describe the probability of an end-to-end distance R using a Gaussian distribution:

$$P(R) = \left(\frac{3}{2\pi \langle R^2 \rangle} \right)^{3/2} e^{-3R^2/2\langle R^2 \rangle} \tag{5.17}$$

where $P(R)$ is the probability that the polymer chain will have an end-to-end distance R. The derivation for this formula is beyond the scope of this book, but can be found in the excellent book on *Polymer Physics* (p. 66) by Rubenstein and Colby.[2] If we find the probability of a particular R using this distribution function, we can find the change in entropy associated with stretching the polymer chain to a different R.

The probability of a particular end-to-end distance for an N-segment polymer ($P(N, R)$) can be expressed as:

$$P(N,R) = \frac{\Omega(N,R)}{\int \Omega(N,R) dR} \tag{5.18}$$

or the total number of possible states (i.e., polymer conformations) with a particular R value, divided by the total number of states with any R. By rearranging this formula for $\Omega(N, R)$, substituting the Gaussian function for $P(N, R)$, and combining this with the definition of entropy, we can obtain:

$$S(N,R) = S(N) - \frac{3k_B R^2}{2Na^2} \tag{5.19}$$

since $\langle R^2 \rangle = Na^2$. The term $S(N)$ is constant for an N-segment polymer and represents the entropy of the unstretched chain. The second term represents the change in entropy as the chain is stretched (i.e., the total decreases).

Using the definition of free energy, we can write an expression for the change in free energy as the chain is stretched to R,

$$dF = -TdS = \frac{3k_B T R^2}{2Na^2} \tag{5.20}$$

The free energy of a system represents the amount of energy available to do work and can be related to an applied extension force by,

$$dF = f dR \tag{5.21}$$

As the chain is stretched, its free energy increases, thus more energy is available to do work. In this case, f is the force applied to produce an end-to-end extension, dR. Extension force is the derivative of the free energy, so

$$f = \frac{dF}{dR} = -\frac{3k_BTR}{Na^2} \qquad (5.22)$$

Here, we can see something quite fascinating. Notice how this entropic spring force is linearly related to the extension; the extended polymer behaves like a classical Hookean spring, generating an opposing force to deformation, proportional to the extension length. The formula can now be seen as a form of Hooke's law, yielding a spring constant, k, of:

$$k = \frac{3k_BT}{Na^2} \qquad (5.23)$$

The spring constant is linearly related to temperature, so a higher temperature will produce a stiffer spring, that is, the polymer will be more difficult to extend at higher temperatures. This intuitively makes sense because the randomizing effects of thermal energy (k_BT) will be more dominant at higher temperatures.

For a more accurate model, taking into account the properties of the solvent surrounding the polymer and adopting a Flory exponent of $v = 3/5$, De Gennes calculated that for an isolated chain the end-to-end distance R is related to the applied force f by,

$$f = \frac{3k_BTR}{R_f^2} \qquad (5.24)$$

where R_f is the Flory radius or the effective unstressed radius of the polymer coil taking into account interactions with the solvent.

5.4.6 POLYELECTROLYTES

A *polyelectrolyte* is a polymer with an ionizable group on the monomer. This means that when the polymer is dissolved in water, the chain will be either positively (cationic) or negatively (anionic) charged. Charged polymers behave a little differently in solution compared to uncharged polymers due to the additional Coulombic interaction between like-charged polymer chains. Within a uniformly charged polymer chain, the monomers should be repelled from each other; hence, charged polymers tend to be more extended in solution because of their increased self-repulsion, although they can also be induced to collapse and even bundle together by the action of additional ions in the solution.

When the thermal energy of a fluctuating polymer chain in solution is on the order of the electrostatic potential energy, we can define a length scale known as the *Bjerrum length* ξ_B. This comparison yields the formula:

$$\xi_B = \frac{e^2}{4\pi\varepsilon_0\varepsilon_r kT} \tag{5.25}$$

The Bjerrum length helps to describe the action of ions in solution around the polyelectrolyte; it approximates a distance within which ions in solution become localized next to the charged polymer. Another useful length scale for ionic interactions with a charged molecule is the *Debye screening length*. This length scale results from an analysis of the electrostatic potential around the charged polymer. If a negatively charged polymer is placed in solution with positive and negative ions (e.g., the Na^+ and Cl^- ions from dissolved sodium chloride), a halo of positive ions will form around the positive backbone. This halo of *counterions* has the effect of screening out a lot of the electrostatic potential from the charged polymer chain. At some distance r from the polymer, the screened electrostatic potential $V(r)$ due to monovalent counterions becomes:

$$V(r) = \frac{e}{4\pi\varepsilon_0\varepsilon_r} e^{\frac{-r}{\xi_D}} \tag{5.26}$$

where ξ_D is the Debye screening length, and ε_r is the relative permittivity of the solution. Beyond this distance from the polymer, the electrostatic potential is effectively zero. This lowering of the effective electrostatic potential from the polyelectrolyte results in reduced interactions between different sections of the polymer chain. Therefore, the chain behaves in a more neutrally charged way and is less extended in solution. This means that adding salt to a polyelectrolye solution will typically reduce the polymer radius of gyration.

5.4.7 Polymer Gels

Filamentous molecules such as polymers can be very long (thousands of monomers) and therefore may become entangled in solution (and possibly cross-linked) to form a gel. Gels are a classic example of a soft material; they are neither solid nor liquid, but somewhere in between. They exhibit no long-range order but do retain the properties of a solid material. For example, many gels can retain their shape unsupported and also exhibit elasticity. Take the gelatin dessert in Figure 5.10. This gel is solid in appearance, and if you deform it (up to a point), it will bounce right back! It is interesting to consider that such a gel actually contains an amazingly small proportion of actual polymer—in this case the gel is more than 90% water. Despite being so dilute, the system retains a stable structure with elastic properties.

FIGURE 5.10 Gelatin-based desserts are an example of a hydrogel. The polymer-like gelatin molecule forms a network structure in water that exhibits relatively solid-like properties. Interestingly, fresh pineapple cannot be added to a gelatin dessert because it contains the enzyme bromelin, which breaks down gelatin; canned pineapple, however, is not a problem as the hot canning process denatures the enzyme.

All gels consist of two different phases; the continuous phase is the solvent in which the "solid" phase is dispersed. There are two different ways in which a polymer gel can be stabilized: by a cross-linking reaction between chains (chemical gels) or by a physical interaction (physical gels).

Chemical gels: Chemical gels are formed when the polymer network is cross-linked by a permanent (or covalent) chemical bond. The strength of this bond means that in general the network cannot be rearranged by heating the sample and thermally breaking the cross-links. An example of a chemical gel is the polyacrylamide gel used in gel electrophoresis (polyacrylamide gel electrophoresis, PAGE) (see Figure 4.22).

Physical gels: Physical gels are not covalently cross-linked and exhibit reversible properties as a function of temperature. A physical gel will "melt" on heating into a liquid-like state known as a *sol*. The gel–sol transition is reversible and can be observed on heating and cooling a physical gel. In gelatin desserts, the gelatin, a filament derived form of collagen, is dissolved in warm water. On cooling, the system undergoes gelation, and hydrogen bonds between filaments form to stabilize the gel. These bonds are relatively weak and heating the dessert will revert it back to a liquid (or sol) state.

5.4.8 HYDROGELS

The term *hydrogel* simply refers to a gel formed with water as the continuous phase component. Hydrogels in an aqueous environment will be swollen with solvent and usually contain a very small amount of polymer compared to their water content. The stability of a hydrogel, however, can be a delicate balance, and many materials exhibit a sudden decrease in volume (or collapse) in response to a stimulus such as pH, temperature, or ionic strength of the solvent.

Hydrogels are found in many applications in food science (Figure 5.10) and health care. In the field of tissue engineering, there is considerable research under way in the formation of polymer networks optimized for the growth of different kinds of cells. It is a challenge for cells to be cultured in a 3D environment, and by introducing them into the 3D scaffold of a hydrogel, the goal of such research is to provide an environment favorable for 3D growth and cell differentiation. Scaffolds currently used in research vary greatly in their composition. They range from constructs based on extracted extracellular matrix components and other natural fiber-based media, to polysaccharides, alginates, and artificially synthesized polymers (Figure 5.11).

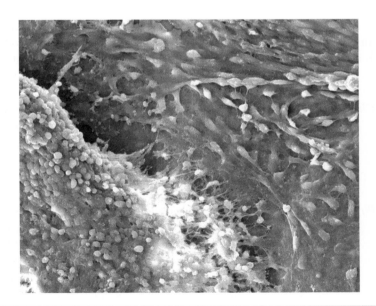

FIGURE 5.11 A scanning electron microscope image of stem cells cultured in a silk-based tissue engineering matrix. (Courtesy of Prof. Wei-Chun Chin, University of California, Merced, CA.)

So-called soft contact lenses are actually made from a polymer hydrogel with a composition of up to about 50% water (the older "hard" lenses were a solid piece of plastic). If you are a contact lens wearer, you will know what happens when you leave a soft lens to dry out; it shrinks and becomes hard like plastic. Dried up lenses will regain their former shape if soaked in water as the polymer network swells and refills with water. Some of the most current contact lens technologies, such as overnight or continuous wear lenses, are based on silicone hydrogels (polymers with a silicon backbone).

5.5   THE GLASSY AND POLYMER MELT PHASES

Up to this point, we have focused on the behavior of polymer chains in a solvent, concentrating on the dilute case, in which the individual chains do not strongly interact with each other. In this section, we will now turn to more concentrated states in polymeric materials and their mechanical properties. Polymers take a variety of forms in everyday objects, but there are really only two significant states that are relevant to our understanding of everyday materials: the glassy state and the melt state. In this section, we are focusing on polymer materials containing very little or no solvent.

When a fluid-like polymer melt is cooled, it goes through the *glass transition* and becomes hard and very useful for manufacturing objects (like the plastic flatware in Figure 5.12). The glass transition is reversible, can be detected by differential scanning calorimetry, and it's characteristic temperature is usually denoted as T_g. Glassy behavior is not just confined to macromolecules such as polymers and is a fairly general phenomenon observed in many amorphous materials. Glasses can be formed by cooling small molecule systems such as ionic salts, organic molecules, and the silicates from which everyday panes of glass are made. Glasses can also be formed in larger-scale materials like colloidal solutions or granular systems.

Glassy behavior in a material is characterized by a very long *relaxation time*. When any soft material is subject to a deformation, the material will respond viscoelastically, then return to an equilibrium state over some characteristic timescale known as the relaxation time, τ. Shearing a polymeric material causes a forced extension of the polymer chains; then as the material relaxes, thermal motions allow the chains to return to their more coiled equilibrium states. The concept of the relaxation time was first introduced by 19th century Scottish physicist James Clerk Maxwell, who is best known for his unification of the electrical and magnetic forces in Maxwell's equations. As a material approaches the glass transition, this relaxation time becomes extremely large, and the material's structure is "frozen in;" it will no longer flow.

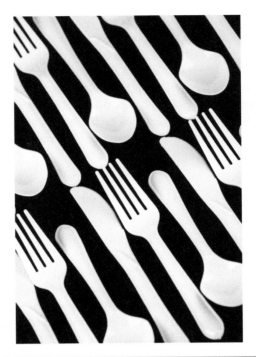

FIGURE 5.12 Polymers below the glass transition T_g are hard. They do not flow and can be molded into durable items.

Below the glass transition, a polymer material will be a strong, hard plastic and can be very brittle. The glassy state is typically amorphous, although some polymers will form crystalline domains throughout the structure where chain segments pack together in an ordered fashion. Domain melting does not occur at the glass transition, but instead these crystalline or semi-crystalline polymers will exhibit a different melting point T_m above the glass transition at which this ordering is lost. The Young's modulus in the glassy state is typically about 10^9 Pa.

The glass transition point does not have an associated latent heat and is not particularly sharp (quite unlike a first order phase transition), it may take place over a range of more than 10°C. You should think of the glass transition more as a change in physical properties (i.e., specific heat capacity, volume) than a true thermodynamic phase transition. The glass transition temperature, T_g, can be determined by two main methods. Differential scanning calorimetry (see Chapter 3 for more details on that technique) will reveal a change in specific heat in the polymer, characterized by a change in the slope of the heat curve. Alternatively, one can look at the rate of volume expansion as a function of temperature, a parameter which will change at T_g

In contrast to the glassy state, polymer chains in the melt state (at temperatures well above the glass transition) exhibit a behavior close to that of the ideal chain. The material is essentially fluid-like with a viscosity η. Polymer melts or rubbers are very deformable, with a Young's modulus of 10^5–10^6 Pa. The name *rubber* originates from the ability of natural rubber to remove pencil marks from the page; however, the extraction of natural rubbers from trees in fact dates back thousands of years, when a thin milky fluid was extracted from the *Hevea brasiliensis* tree in the Amazonian rain forests of Brazil, Bolivia, and Peru. These natural materials originally found many uses as waterproof coatings, although their tendency to melt in the sun and flow restricted their wider use.

To increase the functionality of a polymer melt, the polymer chains can be cross-linked to form a highly deformable elastic material, like the rubber in a balloon (Figures 5.13 and 5.14). The first cross-linked rubber was developed by the U.S. inventor Charles Goodyear in 1839 from natural rubber. Heating the rubber with sulfur created cross-links in a process called *vulcanization*, producing a hard, black material. The discovery of this important process revolutionized the automotive industry, allowing the production of hard-wearing solid rubber tires, an excellent alternative to the wooden wheels of the time!

FIGURE 5.13   Cross-linked polymer melts, such as the latex used in balloons and elastic bands, are soft and elastic.

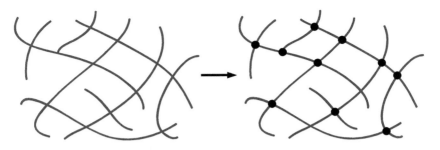

FIGURE 5.14 Cross-linking the entangled polymer chains in a rubber improves durability and elastic properties.

5.6 THE MECHANICAL PROPERTIES OF POLYMERS

A common characteristic of soft materials is that while slow deformations produce viscous behavior, fast deformations may yield an elastic response. Polymers can form materials that are purely elastic, viscoelastic, or viscous, and the behavior of an entangled polymer network is a good example of this phenomenon—the relaxation characteristics depend on the type of mechanical deformation used. For example, on very small length scales, the bonds between atoms in a polymer chain can be slightly stretched. This deformation relaxes back to its equilibrium state quickly. In the same material on a larger, molecular-level length scale, the network itself can be deformed, but in this case will relax back to equilibrium slowly as entangled polymer chains must move around each other in the network.

The length of the polymer chains in a material has a strong impact on viscoelastic properties with a relaxation time τ, related to chain length by the relation,

$$\tau = \tau_0 N^\alpha \tag{5.27}$$

where τ_0 is a time constant, N is the number of segments in the chain, and α is an exponent close to 3. One of the most widely used models for polymer chain behavior in the melt is the *reptation model*. In this model, we imagine the polymer to be confined and fluctuating like a worm within a tube of fixed radius, and an analysis of this model yields an exponent α equal to 3. The significance of the relaxation time here in terms of viscoelastic properties is that chain length is strongly correlated to shear properties. On shear timescales much less than τ, the material will exhibit elastic behavior because of its entangled nature, but on shear timescales much greater than τ, the chains have plenty of time to fluctuate and untangle from each other, creating a flow and thus, viscous behavior.

TABLE 5.1

Examples of the Glass Transition Temperature for Some Different Common Polymers

Polymer	Glass Transition Temperature, T_g (°C)
Polyethylene	−20
Polyethylene terepthalate (PET)	100
Polymethyl methacrylate (PMMA)	105
Polyvinyl chloride (PVC)	82

Note: These values are somewhat approximate as the value of T_g can vary depending on the measurement method used.

The viscoelastic properties of polymeric materials also vary with temperature. Far below the glass transition temperature T_g (see Table 5.1), polymers behave like elastic solids under small deformations. In the glassy phase, polymer chain conformations are effectively frozen in, and the material requires a large force to deform. Think about a piece of hard plastic; it does not seem very "elastic," but plastics, like other solids, including crystalline materials, also obey Hooke's law under small deformations, so:

$$\sigma = Y\varepsilon \tag{5.28}$$

where Y is the Young's modulus as defined in Chapter 2. The strain ε represents the deformation of the material ($\Delta l/l$), and the stress σ indicates how much force is applied (F/A). Close to the glass transition, the properties of a polymer begin to change with an increase in temperature, and the material becomes viscoelastic until some temperature above T_g when the material approaches a predominantly viscous regime.

The viscoelastic properties of a material can be quantified in two different ways using the parameters—*creep compliance* and *stress relaxation*. Creep compliance $J(t)$ is a time-dependent measure of deformation in a material under a constant stress σ_0. Therefore, the quantity depends on the time-dependent strain, $\varepsilon(t)$ according to the equation:

$$J(t) = \frac{\varepsilon(t)}{\sigma_0} \tag{5.29}$$

The creep response of a viscoelastic polymer under a fixed stress can be separated into three different contributions: a fast (ideal) elastic response, a slow elastic response, and a viscous response (or flow). For purely elastic materials, creep compliance is a constant—since they obey the steady state relation, $\sigma = Y\varepsilon$. In a purely viscous material, creep should increase indefinitely as the material flows irreversibly. You can think about creep as the gradual deformation of a normally elastic material under the application of a constant force. For example, a new foam seat cushion will spring back after each time someone sits on it, although over years of use the cushion will gradually deform to a new shape. Any material under constant load, unless perfectly elastic, will be susceptible to creep—gradual deformation with time.

The second important parameter for viscoelasticity is the stress relaxation modulus $G(t)$, given by:

$$G(t) = \frac{\sigma(t)}{\varepsilon_0} \tag{5.30}$$

where $\sigma(t)$ is the time-dependent stress, and ε_0 is a constant applied strain. In this case, if a material was stretched and held in place at a constant strain, the force required to maintain that strain would decrease over time.

One common way to measure the stress relaxation of a polymer is to apply an oscillatory deformation to the material, for example:

$$\varepsilon(t) = \varepsilon_0 \cos(\omega t) \tag{5.31}$$

where $\varepsilon(t)$ is the time-dependent strain. This yields a time-dependent stress, which we can write as,

$$\sigma(t) = \varepsilon_0 [G'(\omega)\cos(\omega t) - G''(\omega)\sin(\omega t)] \tag{5.32}$$

G' is known as the elastic storage modulus, and G'' is a complex number known as the loss (or viscous) modulus.

In any given material, the relaxation moduli will reflect the response of the material across different timescales (e.g., fast or slow deformation). To make a measurement, materials are deformed under a periodic load with frequency ω. Then, G' and G'' are measured across a wide range of frequencies (typically three to four decades). Measurements of $G'(\omega)$ and $G''(\omega)$ can be used to characterize the mechanical properties of soft materials, including polymer networks and colloidal systems. The technique is also known as mechanical spectroscopy. In a viscoelastic material, the elastic modulus will cross over the viscous modulus at the transition point from viscous to elastic bulk behavior and indicates a possible sol–gel transition or the onset of rubbery behavior in a polymer network.

5.7   LIQUID CRYSTAL POLYMERS

As the name implies, liquid crystal polymers (LCPs) are polymeric materials with liquid crystalline properties. We discovered in Chapter 3 that liquid crystal materials are typically composed of small molecules with a stiff, rod-like section. These molecules tend to align to form an anisotropic phase. Polymers may also exhibit liquid crystalline properties, and there are two main classes of LCP (Figure 5.15). Side-chain LCPs have a flexible polymer backbone with rod-like mesogenic side chains, whereas main-chain LCPs consist of a polymer formed from rigid rod-like segments arranged linearly. In both cases, the mesogenic units behave like a nematic liquid crystal in

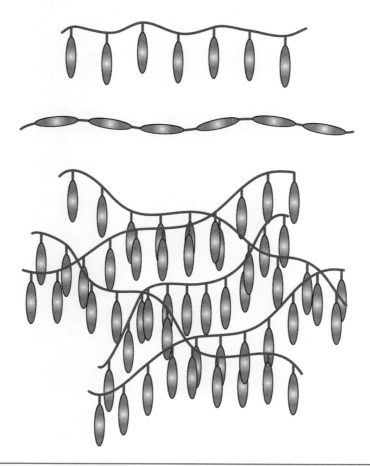

FIGURE 5.15   Liquid crystal polymers can be side chain (top) or main chain (middle). In bulk, the liquid crystalline mesogens can align to form a birefringent phase with nematic-like ordering (bottom).

bulk and align spontaneously to form a liquid crystal phase. Figure 5.16 shows example birefringence textures of an LCP with different alignments with respect to the encapsulating glass plates.

An interesting use of a main-chain LCP is in the hard-wearing material trademarked as Kevlar by DuPont (polyparaphenylene terephthalamide). This material is formed from a main-chain LCP composed of repeated

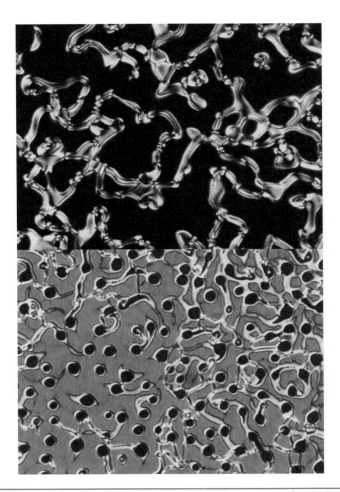

FIGURE 5.16 Polarized optical microscope images (100X magnification) of a side-chain liquid crystal polymer. These images show typical liquid crystalline defect textures for the nematic phase with (top) a *homeotropic* glass treatment, giving a mostly isotropic texture (black) for the nematic phase except around line defects, and (bottom) a *planar* glass treatment. In this image, the material is cooling into the polymer nematic phase from the isotropic liquid phase, and black regions represent material still in the isotropic phase. (Courtesy of Chunhua Wang.)

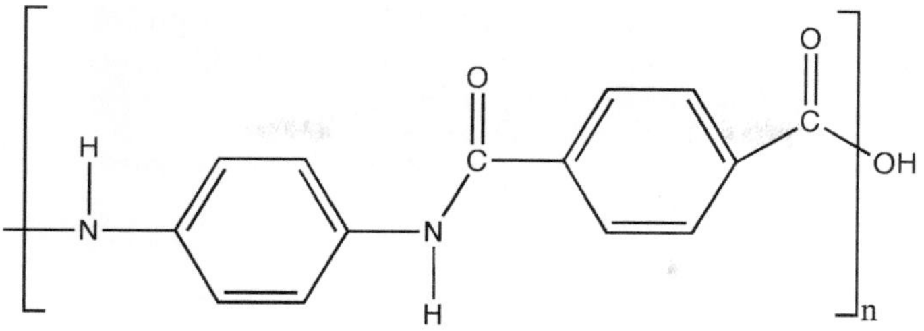

Kevlar

FIGURE 5.17   The molecular structure of Kevlar (polyparaphenylene tere-phthalamide), an extremely strong liquid crystalline polymeric material used in the manufacture of bulletproof vests.

aromatic segments as shown in Figure 5.17 and is a member of a group of materials known as the aromatic polyamides. Note the structural similarities between this monomer and the classic nematic material 5CB (4-cyano-4-n-pentylbiphenyl) shown in Figure 3.6. The polymer has a rigid monomer structure, and as a result the polymer chains have a tendency to align with one another in solution, forming a lyotropic liquid crystal phase. This property means that Kevlar can be easily spun into fibers for use in textile production, and Kevlar is widely used for protective clothing woven from the polymer fibers, including bulletproof vests and other body armor, lighter ropes, and cables, and to increase strength in composite materials. In Kevlar, the spun fibers are particularly strong because hydrogen bonds form between the aligned chains, strengthening the material.

Liquid crystal elastomers represent another class of polymers with liquid crystalline properties. These materials combine the mechanical properties of a cross-linked polymer with the advantages of an ordered liquid crystalline phase.

5.8   EXPERIMENTAL TECHNIQUES

5.8.1   SCATTERING TECHNIQUES

Scattering techniques in soft matter science are extremely important because they allow us to probe the average structure of a material over a wide range of length scales. All weakly ordered materials (i.e., most soft materials) can be investigated using scattering techniques and despite having only short range or even an amorphous structure, there is still plenty of information to be gained from these methods. In Chapter 3, we described

TABLE 5.2 Approximate Wavelength Ranges for Different Scattering Techniques	
Technique	**Length Scale Investigated**
Light scattering	10^{-7}–10^{-6} nm
X-ray scattering	10^{-10}–10^{-7} nm
Neutron scattering	10^{-15}–10^{-6} m

an x-ray scattering experiment on liquid crystal, and in this chapter, we will focus on polymers in solution. In all scattering techniques (see Table 5.2), a material sample is hit with a collimated beam of electromagnetic radiation or particles. This beam scatters from the material as it passes through, and the resulting scattering pattern is measured on a detector. Here, we focus on x-ray scattering for brevity; however, the principles are the same for other electromagnetic wave and particle-based experiments. (In fact, they are equivalent by the de Broglie relation relating wavelength λ to momentum p, $\lambda = \frac{h}{p}$, where h is the Planck constant, $6.626068 \times 10^{-34}\,\text{m}^2\,\text{kg/s}$.)

Some basic scattering theory is reviewed here and in Appendix C, but an excellent resource for specific scattering and diffraction techniques is *Modern X-ray Physics* by Nielsen and McMorrow.[3]

When a material scatters light or x-rays, the incident photons interact with electrons throughout the material. The incident beam can be treated as an oscillating electric field, and this field will induce an oscillation in the electrons it is incident on. As these electrons oscillate (accelerating and decelerating), they will reradiate photons; this is coherently scattered radiation, an elastic process with no energy losses. The intensity I of this scattered radiation as a function of angle θ from the incident axis, a distance r from the material, can be described by,

$$I = I_0 \frac{e^4 \left(1 + \cos^2 2\theta\right)}{m_e^2 r^2 c^4} \qquad (5.33)$$

where e is the electronic charge, m_e is the electron mass, and c is the speed of light. Equation (5.33) correctly describes scattering from a single electron but is therefore not particularly useful for a real bulk sample. Real samples have an electron density distribution and may even incorporate regular structures (i.e., a crystal lattice or a characteristic correlation length) to complicate things further. Regular structures within a material lead to constructive and destructive interference effects between scattered waves, and the pattern formed on a detector by the scattered x-rays will contain rich information about the material's structure. In Appendix C, the scattering theory for a general material is described in a bit more detail, but here we just consider a material without a regular lattice structure. Polymers in solution will not typically have a global crystalline structure but scattering techniques can still be used to quantify their average chain distribution. In Section 5.4, we learned how the structure of polymers in solution can be described using parameters such as the root-mean-squared end-to-end distance and the radius of gyration. Scattering experiments allow us to probe the structure deeply and actually measure the scaling exponents ν introduced in that section.

The structure factor in polymer solutions is a quantity that represents the scattering from the average distribution of the polymer chains as a function of the scattering vector $\mathbf{q}$ and can be represented by,

$$S(\mathbf{q}) = \int \langle \rho(\mathbf{r}) \rangle e^{i\mathbf{q} \cdot \mathbf{r}} d^3\mathbf{r} \tag{5.34}$$

$\rho(\mathbf{r})$ here is the electron density distribution function, representing how the material's mass is spatially arranged and therefore a function of position vector $\mathbf{r}$. $\rho(\mathbf{r})$ captures the distribution of the polymer chains on different length scales (i.e., the chains are extended, swollen, ideal, etc.). In scattering experiments, the function $S(\mathbf{q})$ is the Fourier transform of the real space density distribution (see Appendix A) and therefore provides information on the average distribution of structural length scales in the material. For a solution of polymers, with some characteristic radius of gyration, scattering data can give us a plot something like the cartoon graph for $S(\mathbf{q})$ shown in Figure 5.18. Although this scattering plot does not exhibit any peaks, it still contains useful information. The x axis on the graph in Figure 5.18 represents the scattering vector $\mathbf{q}$. We can relate $\mathbf{q}$ to real space distances by the relation $\mathbf{q} = 2\pi/d$, where d is a real space distance ($\mathbf{q}$ is a vector in reciprocal space). Because of this inverse relationship, on the graph at $\mathbf{q} < P$ we are looking at scattering from large length scales greater than the radius of gyration R_g. Conversely, for values of $\mathbf{q} > P$, we see the results of scattering from length scales less than R_g. The slope of this log-log plot at $\mathbf{q} > P$ gives us D, the fractal dimension (or scaling exponent). D, as we learned in Chapter 2, is a measure of how an object fills space (its fractional dimension) and the scaling exponent for our polymer coil. For a random walk polymer chain, we expect a slope of -2.

Certain polymers are classed as semi-crystalline, above the glass transition temperature short segments of the chain locally crystallize to form ordered aligned domains. For a material with this property, scattering can reveal information on the degree of crystallinity, and lattice parameters of the crystalline domain by comparing Bragg scattering peaks from the crystalline regions and additional amorphous scattering contributions from the non-crystalline regions. In this case, we would expect to see the Bragg peaks superimposed on the typical scattering signature of an amorphous polymer.

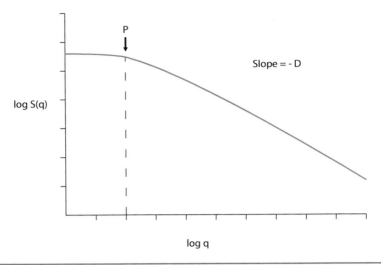

FIGURE 5.18   Schematic graph demonstrating a typical scattering curve from a polymer solution. The structure factor is plotted as a function of q, the scattering vector. Point P indicates a crossover point from scattering representing structural length scales outside the polymer coils (very low q) to inside (higher q). The slope of this log-log plot ($-D$) characterizes the polymer scaling exponent.

5.8.2   POLYMER SPECTROSCOPY

Spectroscopic techniques involve the analysis of electromagnetic radiation as it interacts with a material. As different wavelengths pass through the material, the light can be absorbed or scattered either elastically or inelastically. By analyzing the wavelength distribution of this transmitted or scattered light, a great deal of chemical and physical information can be obtained. In this section, we discuss three different spectroscopic techniques commonly used in polymer science, Fourier transform infrared spectroscopy (FTIR), Raman spectroscopy, and nuclear magnetic resonance (NMR) spectroscopy. Each of these techniques is based on the excitation of different normal modes of vibration within the molecule of study. For polymers, infrared excitation is used with a range of $5 - 100 \times 10^{12}$ Hz. The presence or absence of certain signature modes for various chemical groups can reveal chemical structure, characterize purity, and analyze the composition of polymer mixtures.

5.8.2.1   Fourier Transform Infrared Spectroscopy

FTIR is a technique in which the absorption of different infrared wavelengths of light in a material is measured and therefore can be used to elucidate the chemical structure of the material under investigation. To carry out a measurement, an infrared beam is directed at the sample, and as it passes though, photons of different wavelengths are absorbed. These absorptions take place because different thermal molecular vibrations in the molecule are excited. The molecular bonds in a material have different degrees of freedom; they are not rigid and can stretch and rotate in various ways. The number of these degrees of freedom will vary depending on the chemical structure of the material under investigation. Each different vibrational mode in a molecule has a specific resonant frequency and therefore can be excited if a photon of that frequency is incident and absorbed. The infrared range of the spectrum is a good match for the vibrational frequencies associated with typical modes of bond stretching and rotation.

As the infrared beam passes through the sample, certain frequencies of light will be preferentially absorbed and therefore will be missing from the transmitted spectrum. An analysis of this transmitted spectrum is used to characterize the material.

FTIR spectroscopy makes use of an interferometer to generate a characteristic function representative of the relative intensities of light emerging from the sample. To obtain the spectrum as a function of wavelength, we can compute the Fourier transform of this interference function and obtain full information on wavelength-dependent absorption by the material. Some example data can be seen in Figure 5.19 for three different polymers. This method is widely used and a considerable advantage over earlier methods that used a much slower wavelength scanning technique.

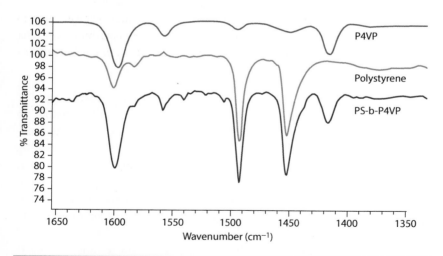

FIGURE 5.19 An example of an FTIR spectrum for three different example polymers: poly(vinyl pyridine) (P4VP), polystyrene, and a poly(vinyl pyridine)–block–polystyrene (PS-b-P4VP). In the data shown, we can clearly see that the block copolymer has characteristics of both of its individual polymer block components. The C = C stretch mode is visible at about 1500 cm^{-1} for all three molecules. (Courtesy of Chai Lor.)

5.8.2.2 Raman Spectroscopy

The Raman spectroscopic technique relies on the abilities of a material to scatter light inelastically. Raman scattering, occurs when a photon excites a thermal mode of vibration in a molecule similar to the vibrations probed in infrared spectroscopy. For example, the excited mode could be a bond stretch or a rotation. The difference in this technique is that instead of the photon being fully absorbed, it is absorbed and then reemitted at a lower frequency. The change in frequency of the detected photon is equal to the resonant frequency of the excited vibrational mode.

Raman spectroscopy data are usually plotted as spectra in terms of the wave number k (or Raman shift), measured in cm^{-1}. The typical range of wave numbers probed is between 200 and 4000 cm^{-1}. For example, the OH "stretch" mode will have a wave number of 3400 cm^{-1}, and the CH stretch has a wave number of 2290 cm^{-1}. To take a measurement, the wavelength intensity at different frequencies is collected and plotted as a function of the wave number.

Raman scattering is actually a very weak effect, producing low scattering intensities. If we compare Raman with the strong Rayleigh scattering process, the Raman signal can be 10^3–10^6 times less intense. This weak signal can be a problem experimentally, so to maximize intensity a laser is typically used for excitation. Filters are also employed to cut out the intense Rayleigh scattering signal (which does not change in frequency). Sample fluorescence can be another problem for detection of weak Raman signatures. Fluorescence is a different process in which incident photons are absorbed and reemitted at lower energies by the material. Fluorescence effects can be minimized by bleaching (exposing the sample to a high-intensity light) before the measurement is carried out.

5.8.2.3 Nuclear Magnetic Resonance

NMR is an important experimental tool for polymer science and is used to study molecular structure and dynamics. The technique is a key method in the design of new polymers and can be used to identify atoms present, functional groups, and their configurations. NMR is also useful for measuring the average molar mass of a material, molecular tumbling correlation times, and other localized dynamics.

The NMR technique was first discovered in the 1940s and takes advantage of the behavior of nuclear spins in a material when subjected to an applied magnetic field. All nuclei have a property called "spin" associated with them. In a simplistic model, you could think of the atomic nucleus as a rotating body with some angular momentum. This angular momentum is quantized and therefore can only take certain values; these values are the nuclear spin. In NMR techniques, we focus on the atoms in a material with a nuclear spin of ½. These are atoms with nuclei with odd mass numbers, such as H^1 or C^{13}, and have half-integral spin quantum numbers (m).

The nucleus is analogous to a tiny bar magnet and so has an associated magnetic moment μ. When placed in an external magnetic field H, the nuclear spins tend to line up either parallel to the field direction or in the opposite direction (antiparallel). These two spin states have different associated energies separated by:

$$\Delta E = 2\mu H \tag{5.35}$$

The lower energy state, parallel to the field, is slightly more likely to occur, and there will always be a small temperature-dependent excess population of nuclei in the parallel state. This energy level splitting in an applied magnetic field is known as *Zeeman splitting*. If we wanted to increase the population of antiparallel nuclei in the sample, we would need to excite some of the parallel nuclei from the lower energy level by increasing their energies by ΔE. The energy difference between the two different spin states can be expressed as a frequency v:

$$v = \frac{2\mu H}{h} \tag{5.36}$$

where h is the Planck constant. NMR involves exciting parallel nuclei into the antiparallel state, then measuring their relaxation back to equilibrium (see Figure 5.20). The technique is very sensitive to molecular structure and conformation because the atomic environment around any particular atom modifies the magnetic field experienced by the nucleus, therefore producing a characteristic resonant frequency.

There are two important steps to the measurement of an NMR signal:

1. The sample must be placed in a high magnetic field. In a solution sample, initially all the atoms are randomly oriented with no net magnetic moment, but when an external magnetic field (typically a high field of several tesla) is applied, they become aligned with the field in either the parallel or the antiparallel state. Because there is an excess population in the parallel state, the sample has an overall magnetic polarization

2. The sample is excited using a radio frequency. The sample is pulsed with a radio signal across a range of frequencies, exciting all spins simultaneously. This signal excites parallel spins into the higher-energy antiparallel state, then these excited spins decay back to their lower-energy equilibrium state, emitting radio-frequency photons for detection. Calculating the Fourier transform (see Appendix A) of this signal will recover the NMR spectrum.

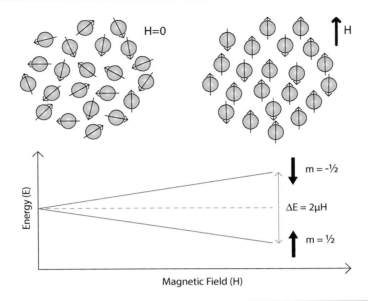

FIGURE 5.20 Before magnetic field application, the nuclear spins are randomly distributed in the sample, but after field application they become aligned either parallel to the H field direction (the majority) or antiparallel (the minority). The graph at the bottom shows the energy difference between the two spin states as a function of H. The two states are assigned the magnetic quantum numbers $m = -\frac{1}{2}$ and $\frac{1}{2}$.

A key feature of NMR spectroscopy is that the magnetic field to which a nucleus is exposed is not just equal to the applied field but is modified by the chemical environment of the nucleus. This results in a spectrum of resonance peaks for any given material. The nucleus of any atom is surrounded by negatively charged electrons that provide some shielding from the applied field. Different atomic bonds modify the shape of the electron cloud, resulting in different effective magnetic fields for different chemical groups and bonds; therefore, a spectrum of different resonant frequencies can be observed. Another modification to the resonant frequency for a particular nucleus in a material comes from interactions between nuclei connected through one to three covalent bonds and the resulting modification of the local magnetic field. This effect is called spin–spin coupling and results in split peaks that can be interpreted as corresponding to different atomic bonds and are used to assign spectra. A three-bond coupling can be used to estimate the intervening torsion angle.

The resonant frequency for a particular nucleus (say ^{1}H) depends on the applied magnetic field for the instrument; so, these frequencies must be expressed as shifts from a reference material, a compound added to the sample with a clear, sharp NMR signal is used, typically tetramethylsilane (TMS). The position of the NMR signal relative to the reference sample for a particular nucleus is known as the *chemical shift*.

NMR experiments usually detect the ^{1}H signal, and protonated solvents produce strong signals that obscure nearby, usually more interesting, ^{1}H signals. A common solution is to use deuterated solvents. Deuterium (^{2}H) has a resonance frequency that is about 3/20 of that of the proton and therefore does not interfere with ^{1}H signals. In addition, for a molecule with significant signal crowding and overlap, deuteration of specific sites within the molecule can simplify the ^{1}H spectra considerably.

QUESTIONS

Polymer Architecture

1. Explain the following terms in your own words: (a) homopolymer, (b) tacticity, (c) dispersity, and (d) oligomer.
2. A polymer has the following molecular mass distribution. Calculate the weight averaged molecular mass, the number averaged molecular mass, and the dispersity index of the polymer.

Percent of Composition (%)	Mass (Da)
2	1000
5	2000
10	2500
60	3000
20	4200
2	5000
1	8000

3. Explain the difference between a homopolymer, a random copolymer, and a block copolymer.
4. A random copolymer is composed of 20% polyethylene and 80% polyvinyl chloride monomers. Make an estimate for the average molecular weight if the average molecule is composed of 1000 monomers.
5. What does it mean if the dispersity index, Đ of a polymer is equal to one? Can Đ ever be less than one?

Polymers in Solution

6. What is the average volume filled by a 10,000 Da PEG molecule in a theta solvent?

7. A polymer has a molar mass $M = 86$ g/mol and is composed of 1000 identical monomers, each 2 nm in length. Calculate the root-mean-squared end-to-end distance and estimate the concentration at which the solution enters the semi-dilute regime

8. Consider a one-dimensional polymer of length N monomers lying along the x axis. Each successive segment (of length a) can point either in the $+x$ or $-x$ direction with equal probability

   a. What is the average and mean-squared end-to-end dizstance?

   b. In the presence of a field, such as a flow, assume that the probabilities to point in the $+x$ and $-x$ directions are p and $1-p$, respectively ($p > \frac{1}{2}$). What are the average and mean-squared end-to-end distances in this case?

   c. Comparing your answers to a and b, what can you say about the polymer's response to flow?

9. In many cases, polymers have to squeeze their way through tight spaces, especially biopolymers in the cellular context. When a polymer passes through a narrow channel, it is favorable for the molecule to adopt the configuration of a chain of "blobs," each with freely jointed chain behavior.

   Consider a freely jointed chain of length N monomers with monomer length a that is trying to squeeze through a cylindrical pore of diameter D and length L. Assume $D \gg a$ and $L \gg Na$. Very small sections of the chain are unlikely to be affected by the confinement. As the length of this section grows, you will reach a point at which confinement affects the configuration of the polymer. In the following questions, do not worry about exact factors and simply give the scaling forms of the quantities

   a. What is the volume occupied by a blob in terms of the parameters defined?

   b. How many monomers does each blob contain, assuming self-avoiding chain behavior inside the blobs?

   c. Find an expression for the free energy of a blob.

   d. The entire chain can be thought of as a string of blobs. How many blobs are there? How much total free energy? This represents the entropic cost of squeezing the polymer into this pore.

10. Do long and short polymers have the same density in solution? Can they be sorted by centrifugation?

11. As temperature increases, will a single ideal polymer chain become more or less elastic?

12. I take a thick elastic band and stretch it as far as it will go. Then I release the band suddenly and let it snap back. The band feels warm—can you explain why?

13. The probability of an ideal polymer chain having an end-to-end distance R is given by the formula:

$$P(R) = \left(\frac{3}{2\pi\langle\overline{R^2}\rangle}\right)^{3/2} e^{-3R^2/2\langle\overline{R^2}\rangle}$$

a. Explain qualitatively how stretching the chain will decrease its entropy.

b. Show that after stretching the chain to an end-to-end distance R_1, the total entropy is given by:

$$S(N,R_1) = S(N) - \frac{3k_B R_1^2}{2Na^2},$$

and that the term $S(N)$ is not a function of R and therefore a constant for a fixed chain length of N segments.

Experimental Methods

14. The structure of a dilute solution of polymer molecules is investigated in an x-ray scattering experiment. If the molecules are dissolved in a theta solvent, what will the expected slope of the $S(q)$ graph be on a log–log plot? How will this slope change if the polymer is swollen or collapsed?

15. What is the significance of the crossover point in a rheology measurement where G' is equal to G''?

16. The Raman wave number for the CH stretch is 2290 cm⁻¹. For this vibration, calculate (a) the wavelength and (b) the vibrational energy in electron volts.

REFERENCES

1. T.A. Witten and P.A. Pincus, *Structured Fluids: Polymers, Colloids, Surfactants.* Oxford, UK: Oxford University Press (2004).
2. M. Rubinstein and R.H. Colby, *Polymer Physics (Chemistry).* Oxford, UK: Oxford University Press (2003).
3. J. Als-Nielsen and D. McMorrow, *Elements of Modern X-ray Physics.* New York: Wiley (2001).

FURTHER READING

D.I. Bower, *An Introduction to Polymer Physics.* Cambridge, UK: Cambridge University Press (2002).
P.G. de Gennes, *Scaling Concepts in Polymer Physics.* Ithaca, NY: Cornell University Press (1979).

A.M. Donald, A.H. Windle, and S. Hanna, *Liquid Crystalline Polymers* (*Cambridge Solid State Science*), 2nd ed. Cambridge, UK: Cambridge University Press (2006).

P. Flory, *Principles of Polymer Chemistry.* Ithaca, NY: Cornell University Press (1971).

Y. Osada and A. Khokhlov, *Polymer Gels and Networks.* Boca Raton, FL: CRC Press (2001).

G. Patterson, *Physical Chemistry of Macromolecules.* Boca Raton, FL: CRC Press (2007).

G. Strobl, *The Physics of Polymers: Concepts for Understanding Their Structures and Behavior,* 3rd ed. New York: Springer (2007).

I. Teraoka, *Polymer Solutions, an Introduction to Physical Properties.* New York: Wiley (2002).

Colloidal Materials

6.1 INTRODUCTION

The subject of this chapter represents one of the most diverse areas in soft matter science. Colloidal materials are systems in which small droplets or particles of one material are dispersed in a continuous phase of another material. This definition is deliberately broad because colloidal systems span an extremely wide range of materials, from solid particles suspended in aqueous solution, to droplets of moisture in the air (Figure 6.1), foams, and can even be extended to include granular materials like sand. Colloidal science is a subject that is particularly relevant to our everyday lives because it plays an important role in the manufacture of numerous everyday substances. Many of the foods we eat can be described as colloidal systems. For example, creamy foods, like mayonnaise, sauces, or ice cream, contain tiny droplets of fat dispersed in an aqueous medium (liquid or solid). Food can also be a liquid foam like whipped cream, or a solid foam (like bread or cake). Personal care products like face creams and toothpastes are based on colloidal systems, as are household paints and inks (Figure 6.2). Even the dilute polymer solutions discussed in Chapter 5 can be considered colloids, where long chain polymers adopt a globular conformation in solution.

FIGURE 6.1 A foggy day on campus at the University of California, Merced. Fog is a type of colloidal system, a liquid aerosol of water droplets in air.

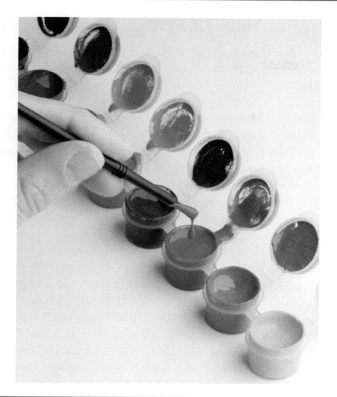

FIGURE 6.2 Emulsion paints are composed of a colloidal dispersion of pigment particles in a water-based medium. The different colors in emulsion paints come from the specific spectral absorption and emission wavelengths of the pigments used. Surfactants are used to stabilize the dispersion.

You may be most familiar with the concept of colloids as applied to solid particles dispersed in a liquid medium, and for much of this chapter we will consider our colloidal materials to represent this common kind of system. It is important to realize right from the start, however, that colloidal systems can be much more broadly defined, and in the topics covered further in this chapter, we look at a range of different colloidal materials. Colloidal systems have a very large surface area-to-volume ratio; therefore, surface properties are an extremely important aspect of colloidal science. The balance of forces between particles and the solvent defines the structure of colloids. Throughout this chapter, we will focus on the interactions between dispersed colloidal particles in a fluid medium, their attractive and repulsive potentials, and the effects of temperature on dispersion and aggregation.

6.2 CHARACTERISTICS OF COLLOIDAL SYSTEMS

Although the classification of colloids covers an extremely diverse array of materials, in general, colloidal systems can be identified by the following characteristics:

1. Droplets or particles of one phase are dispersed in a continuous phase of another material
2. The particles in the dispersed phase have a size between about 10 nm and about 10 μm
3. The dispersed material is characterized by a very large surface area-to-volume ratio, resulting in a large interfacial area between the two components
4. The energy of interparticle interactions between particles is close to the thermal energy $k_B T$

In Table 6.1, a selection of colloidal systems is categorized by combinations of the dispersed and continuous phase. This table should give you an idea of the diversity of this type of material. The word *colloid* was first coined by the chemist Thomas Graham in 1861 from the Greek word for "glue." Colloidal particles tend to stick to each other unless mediated by another force, so this name is appropriate (Figure 6.3). Incidentally, Graham also introduced the words *gel* and *sol* to the scientific vocabulary. A *sol* is a dilute dispersion of solid colloidal particles in a liquid phase, whereas a *gel* in the colloidal sense is a concentrated dispersion that does not flow under low applied stress. Colloidal gels are analogous to the polymer gels we described in Chapter 5 and colloidal suspensions are also similar in

TABLE 6.1
Examples of Different Colloidal Systems[a]

		Continuous Phase		
		Gas	**Liquid**	**Solid**
Dispersed Phase	**Gas**		*Liquid foams:* Soap suds, Cappuccino froth	*Solid foams:* Pillow foam, Styrofoam packaging
	Liquid	*Liquid aerosols* Fog, Hair spray	Mayonnaise, Milk	Ice cream
	Solid	*Solid aerosols* Smoke	Paints, muddy river water, Blood	

[a] Colloidal systems are composed of a dispersed phase and a continuous phase. This table summarizes a variety of examples.

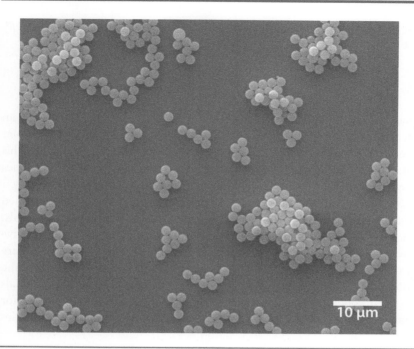

FIGURE 6.3   The competition between interparticle forces and thermal motion determines whether colloids will cluster together in solution or disperse. These 2-μm polystyrene beads have aggregated together as a result of a net attractive force between them. This image was acquired by then drying the particles clusters onto a smooth substrate and imaging using scanning electron microscopy.

their time-dependent behavior to polymer solutions. You will recognize some of the equations we use here from that chapter.

6.3   COLLOIDS IN SUSPENSION

If a material is composed of small particles of one phase dispersed in another fluid phase, the particles may be subject to significant Brownian motion. The extent of this motion depends on particle size. In Section 1.4, we introduced the concept of Brownian motion for a small molecule or particle dispersed in a continuous liquid or gas phase. In a colloidal system, this description directly applies, and we can apply Brownian motion to the thermal motion of colloidal particles in a fluid. In general, colloidal particles are small, and the length scale, λ, of their *random walk* depends on their radius, R, temperature of the system, T, and the viscosity, η, of the suspending medium. In 3-D,

$$\lambda = \sqrt{6Dt} \tag{6.1}$$

The Stokes-Einstein equation can be used to provide a good estimate for D, the diffusion constant for our particle.

$$D = \frac{k_B T}{6\pi\eta R} \tag{6.2}$$

where k is the Boltzmann constant. Using Equations (6.1) and (6.2), we can quickly estimate how far we expect particles to diffuse out from a central point over time. It's important to remember here that λ is an average, and that this calculation is based on the idea that we have a very large number of particles in our system. The distance travelled by each individual particle varies greatly.

In a colloidal system, the overall structure of the material is determined by the interaction between thermal fluctuations and interparticle potentials. Thermal fluctuations promote dispersion, whereas gravity and attractive interparticle forces promote separation or aggregation. The temperature of the material is critical in determining equilibrium structure because the potentials responsible for stabilizing the material are about $k_B T$. Potentials between colloidal particles may be attractive (van der Waals forces, for example) or repulsive (Coulomb repulsion). Depending on the magnitudes

---

WORKED EXAMPLE 6: DIFFUSION

How far, on average, can you expect a 100 nm latex particle to diffuse in room temperature water, in 1 minute?

Assuming that the particle is spherical, with radius, R, and its diffusion in water is governed by the Stokes-Einstein equation, we can write the diffusion length, λ, as:

$$\lambda = \sqrt{\frac{k_B T t}{\pi\eta R}}.$$

Next, we need to find the viscosity of water at room temperature, we can use a value of 8.90×10^{-4} Pa · s at 25°C (298 K). For 1 second of diffusion, this gives us:

$$\sqrt{\frac{(1.38\times10^{-23})(298)(1)}{\pi(8.90\times10^{-4})(100\times10^{-9})}} = 3.8 \; \mu m.$$

The 100 nm particles in water at room temperature will diffuse an average distance of about 3.8 μm every second.

of the various attractive and repulsive forces present between particles, thermal motion may be enough to keep the particles in suspension. If not, then the particles will aggregate together and separate out of suspension. In the following sections, we look at the origins of some different attractive and repulsive interparticle forces that are particularly important for colloidal systems.

6.4   COMPETING FORCES IN COLLOIDAL DISPERSIONS

Colloidal systems are stabilized by a delicate balance of competing forces. These forces either act to repel the colloidal particles from each other or mutually attract them into aggregates. Surface forces tend to dominate because colloids are typically small and can have an extremely high surface area-to-volume ratio. Take for example a 1-cm³ solid cube of gold; the surface area-to-volume ratio of this cube is 6 cm²/1 cm³. Now, if we take that same cube of gold, but split it into 10-nm wide cubes, the total surface area-to-volume ratio of the system increases dramatically to 6×10^6 cm²/1 cm³. This calculation demonstrates the increasingly important role of surface forces as particle size decreases.

The role of the gravitational force cannot be ignored when we think about colloidal stability. In a fluid containing dispersed colloids, competing forces act on the particles. Gravity promotes particle separation by density variation, and interparticle forces promote either aggregation or dispersion. Over time, if particles are large enough, they will gradually sink (*sediment*) or float (*cream*) at a rate that depends on their density with respect to the medium. The effects of gravity are counterbalanced by the thermal (Brownian) motion of the particles and subsequent drag forces as they move around in the fluid. The effects of Brownian motion can be significant enough such that no gravitational separation is observable over long timescales.

Let's consider a spherical particle of radius R submerged in water. The particle has a higher density than water and thus a tendency to sediment. By Archimedes' principle, the buoyant force F_b on the particle is equal to the weight of water displaced, so:

$$F_b = \rho_w g \frac{4}{3} \pi R^3 \tag{6.3}$$

where ρ_w is the density of water (1000 kg/m³), and g is the acceleration due to gravity, 9.81 m/s². In addition to the gravitational force, F_g, the submerged particles are subject to a drag force F_D opposing particle motion as they move at velocity v through a liquid phase with viscosity η. We can approximate F_D using Stokes' drag for a spherical particle,

$$F_D = 6\pi\eta R v \tag{6.4}$$

The net force, is therefore equal to:

$$m\frac{dv}{dt} = F_g - F_B - F_D \tag{6.5}$$

and we can predict a terminal sedimentation velocity if this equation is equal to zero. Diffusion works to counteract sedimentation because in a large system, particles tend to diffuse from areas of high concentration to areas of low concentration (until the system reaches equilibrium at a uniform particle distribution), following the mass transport equation,

$$\frac{dm}{dt} = -DA\frac{dc}{dx} \tag{6.6}$$

where dc/dx is the local concentration gradient in the direction of transport, A is the cross-sectional area perpendicular to the direction of transport, and the diffusion coefficient D is a measure of the mean-squared displacement of a particle per unit time. We can approximate our diffusion coefficient for a spherical particle using the Equation (6.2).

$$D = \frac{k_B T}{6\pi\eta R}.$$

If sedimentation is slow compared to diffusion, then the particles will tend to remain in suspension. For example, when the particles are small or the suspending fluid has a low viscosity, the diffusion rate is high. These equations also demonstrate the importance of temperature in colloidal sedimentation; at higher temperatures, diffusion increases, promoting stability of the suspension.

6.5 INTERPARTICLE INTERACTIONS

In Chapter 2, we introduced several intermolecular forces that can play an important role in the stability of soft matter systems; these include van der Waals attraction, electrostatic repulsion and attraction, and hard sphere repulsion. Such forces are important in colloidal suspensions, and we briefly review these forces in the following section in the context of colloidal particles. In addition, we also introduce two additional relevant interactions: depletion forces and steric repulsion. It should be noted here that all of the forces relevant to colloids are long range and essentially derive from electrostatics; therefore, they can be understood using the principles of classical physics.

6.5.1 VAN DER WAALS ATTRACTION

The van der Waals attractive potential derives from a dipole–dipole interaction between the two neutrally charged particle surfaces separated by a distance r (see Section 1.3.1), where $U \sim \frac{1}{r^6}$ and provides a relatively short-range force. In the absence of any other forces, therefore, we should expect all colloidal particles to gradually stick to each other and aggregate over time. In reality, because of the short-range nature of this force, other interactions in the system often tend to dominate, and by controlling the strength of these interactions colloidal suspensions can be stabilized under different conditions despite the strong short-range van der Waals attraction.

6.5.2 ELECTROSTATIC FORCES

The surface charges on similar colloidal particles can provide a strong repulsive potential, where Coulomb forces act to disperse the particles. If, however, the solution in which the particles are dispersed contains oppositely charged ions to those on the particle surface, then the electrostatic interaction between particles will be modified by the dissolved ions. The resulting electrostatic screening effects can greatly reduce the interparticle Coulomb potential, potentially allowing the particles to even stick together.

When charged particles are placed in an ionic solution, a halo of oppositely charged ions will form around the charged particle. This is known as the *electric double layer*. More details on this phenomenon are included in the section in this chapter on measuring the zeta potential (Section 6.10.2). For monovalent ions, this ionic halo effectively screens the surface charge of the particle for distances greater than a characteristic length we call the *Debye screening length*. The Debye screening length, λ_D, depends on both the concentration of charged ions in the solution n_0 and their valency z according to,

$$\lambda_D = \left(\frac{\varepsilon \varepsilon_0 k_B T}{2e^2 n_0 z^2} \right)^{1/2}$$

$$(6.7)$$

As you can see in Equation (6.7), multivalent ions will have a much stronger screening effect on charged colloidal particles, and the Debye length decreases with increasing valency. Adding highly charged ions to a solution can rapidly induce aggregation of the dispersed particles, so modification of electrostatic particle interactions by dissolved ions can be a powerful driver of colloidal phase behavior. The screening effect of the electric double layer effectively turns the Coulomb repulsion between charged particles into a relatively short-range force. In the absence of a long-range Coulomb repulsion, thermally fluctuating particles are more likely to come close enough to be attracted via van der Waals forces and aggregate.

An interesting example of electrostatic forces inducing aggregation can be seen in the deposition of silt at the mouth of a river delta. One of the mechanisms driving this process derives from the change in ionic concentration as the water turns from fresh river water to salty seawater. Muddy river water contains a large number of tiny suspended clay particles. As the river approaches the ocean, there will be a transition from the freshwater of the river, containing few charged ions, to salty seawater. Seawater contains charged ions, and these ions create an electric double layer around the charged clay particles, screening their repulsive potential. The particles can then aggregate and become larger; at some size limit, they are no longer stable in suspension and precipitate out of solution to become silt at the river mouth.

6.5.3 DLVO THEORY

In the 1940s, a scientific theory describing the delicate balance of colloidal dispersion and aggregation was developed by Derjaguin and Landau[1] and by Verwey and Overbeek.[2] The theory is usually called the DLVO theory for short. In this theoretical description, a potential V is used to describe the different interactions between colloidal particles in solution and we can simply write:

$$V = V_a + V_r \qquad (6.8)$$

for this interaction potential, where V_a is an attractive potential between particles, and V_r is a repulsive potential. A third potential representing the role of the solvent can also be introduced into this formula. Contributions to the potential originate primarily from:

1. Repulsion of the electric double layers
2. Van der Waals attraction between the particles

The theory predicts that for charged colloids in solution at low salt concentrations, there will be a strong repulsion between particles, resulting in a stable colloidal suspension. When more salt is added to the system,

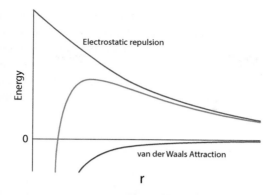

FIGURE 6.4 An example of a DLVO curve plotting potential energy as a function of distance, r in green as the sum of van der Waals attraction and the electrostatic repulsive forces. In some cases, the green curve can dip below the zero potential line, resulting in a second stable minimum at which particles can be weakly bound to each other.

the more likely the particles will be to aggregate. We can see an example of this potential in Figure 6.4.

The predictions of DLVO theory describe the behavior of a simplified colloidal solution and should predict if a colloid will aggregate or remain suspended in a given solvent; however, in real systems this simplified model is not often accurate. For example, these models do account for the effects of steric interactions between particles or excluded volume effects. These effects are discussed in the following sections. Other factors not considered include the effects of water structure, particle hydrophobicity, and any specific ionic arrangements around the particles. DLVO theory is a good starting point to think about the behavior of colloidal solutions, but in most real systems there are other complicating factors that must be taken into account when determining the precise conditions for aggregation and dispersion.

6.5.4 DEPLETION FORCES

The van der Waals attraction is not the only driver of aggregation for neutrally charged colloidal particles; a reversible attractive force can be generated from what is known as *depletion attraction*. The effect can be produced if a second, smaller particle, is mixed in with larger colloidal particles in a solution, leading to aggregation of the larger particles. The mechanism for this net attractive force between the larger colloids can be easily explained qualitatively; its origin lies in the thermal motion of the smaller particles in the system.

As we have learned, all colloidal particles in a fluid experience Brownian motion as a result of the large number of collisions they experience with other particles and solvent molecules. If we first consider a single large colloidal particle, surrounded by smaller particles (they could be solvent molecules or smaller colloids), this larger particle at temperature T will be subjected to constant bombardment from all sides. On average, over time the net force is isotropic, as the smaller particles hit the large particle from all directions with equal frequency. Now expand this idea to a solution with two different particle sizes, as shown in Figure 6.5. The large colloids, when separate, move with a

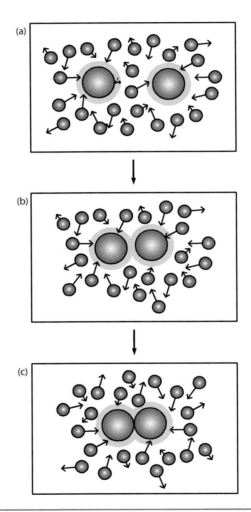

FIGURE 6.5 An attractive force between neutrally charged colloids results from the depletion effect. (a) The two larger colloids are far apart and bombarded by smaller particles from all sides. (b) The large particles approach each other with a separation less than r (the radius of a smaller particle), then (c) the small particles can now only push the large particles together.

random walk, until at some time they happen to closely approach each other. As the large colloids approach closer than a separation of about $2r$ (where r is the radius of the smaller particles), it becomes increasingly difficult for the smaller particles to drift between them. In this situation, the random forces on the larger colloids are no longer isotropic because no smaller particles can get in between and produce collisions to drive them apart. Consequently, the net Brownian forces push the larger colloids together, and once in this state, they tend to stay together.

This net force is known as a depletion force, and by tuning the size of the smaller particles in the solution, it is possible to control the aggregation of the larger particles. Note that the depletion attraction does not necessarily result in permanently aggregated particles (as we would expect for van der Waals attraction), and by changing solution conditions it is possible to redisperse the particles. The effects of this depletion force per unit area are sometimes referred to as the osmotic pressure Π.

Depletion attraction effects can also be induced by adding polymer molecules to a colloidal solution, provided the polymer is not attracted to the colloidal surface. As we explored in Chapter 5, polymers in a dilute solution will adopt a globular random walk configuration, characterized by R_g, the radius of gyration. Free in solution, it is energetically unfavorable for the polymer to extend much beyond R_g; therefore, we can think about the polymer as analogous to a colloidal particle of radius R_g. Once the larger colloids approach each other with a separation less than $2R_g$, they experience a net force toward each other due to the surrounding polymer globules and tend to stick together. One difference in using polymers instead of hard particles to induce this attractive effect is that the radius of gyration for a polymer is tunable with temperature (R_g is reduced at higher temperatures). This means that small changes in temperature can be used to control the particle dispersal. Another important factor to take into consideration is that although the polymer chains are relatively globular when unconfined, they can adopt different conformational shapes in a confined space. Extension of polymer "particles" may allow them to squeeze between the colloidal particles, whereas a solid particle cannot.

6.5.5 STERIC REPULSION

One mechanism for controlling the attraction of colloidal particles to each other is by *steric* modification of the particle surface (i.e., changing the morphology of the particle). To achieve an effective repulsion between particles, the surface can be coated with a polymer. Picture the "hairy particles" depicted in Figure 6.6. When two polymer-coated surfaces approach each other, their flexible chains start to interact. For the particle surfaces

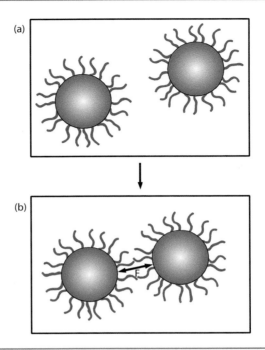

FIGURE 6.6   (a) Polymer brushes on the surface of colloids result in a steric repulsion (b) between the particles when they approach each other.

to approach closely, these chains must interdigitate. The greater the interdigitation of the adjacent polymer shells, the greater the steric repulsion between particles. So, what is the fundamental origin of this steric repulsive force? On the surfaces of the two approaching particles, the long polymer chains thermally fluctuate, constantly varying in conformation similar to the free polymer chains discussed in Chapter 5. These attached chains exhibit an effective radius of gyration depending on their surface density (the chain is not completely free because it is anchored at one end to the particle). A dense coating of polymers on the colloidal surface will result in more extended chains forming a "comb" or "brush" configuration due to lateral interactions between the polymer chains (Figure 6.6a). Imagine bringing two of these polymer brush-coated colloids close together to try to interlock the fluctuating polymer chains (Figure 6.6b). Intuitively, you can see that it is unlikely that they will interlock; there is a large energy barrier to this arrangement. For the chains to interlock with each other, there would need to be a considerable decrease in the entropy of the system because in the interlocked configuration the chains would be confined to a restricted number of possible chain conformations. We can relate this idea to the concept of the entropic spring introduced in Chapter 5. If the number of possible states (in this case chain conformations) in a system is reduced, then the entropy is decreased. Since systems tend to be found in their highest entropy state, it is highly unlikely that a close approach of two particles will occur, as this would require spontaneous interdigitation of the polymer chains on the two particles in a quite specific configuration. The closer the approach, the greater the drop in entropy, and therefore the greater the effective repulsive force.

The concept of the polymer brush coating can be applied generally to surfaces as well as colloidal particles, rendering them resistant to the absorption of large molecules, such as proteins. Polyethylene glycol (PEG) is an example of a commonly used hydrophilic polymer that is often applied for this purpose, although the choice of molecule depends on the application of the particles. PEG molecules can be conjugated to a surface-binding chemical group to create a uniform brush phase on the surface of a substrate. For example, an SiO_2 surface can be easily coated with a molecule such as a PEG-silane molecule (shown schematically in Figure 6.7). By coating surfaces with a polymer brush arrangement, antifouling surfaces can be designed to resist protein absorption for biomedical implants or for self-cleaning industrial applications.

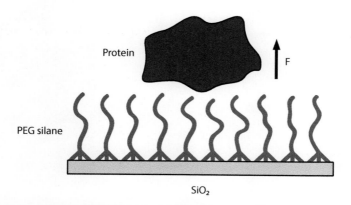

FIGURE 6.7   Surface modification using polymers in the "brush" phase can be used to reduce protein absorption in surfaces by steric repulsion. As particles or proteins approach the treated surface, they will experience a repulsive force, F, which is entropic in origin. This technique is useful in the treatment of biomedical implants. Protein fouling can shorten the life of an implant considerably, but polymer coatings reduce this problem by slowing the rate of non-specific surface absorption.

6.6   COLLOIDAL AGGREGATION

When colloidal particles in solution are subject to a net attractive force, they begin to aggregate into clusters. As long as these clusters are small, they continue to diffuse at a rate dependent on their size, as we described earlier using the Stokes-Einstein equation (see Section 6.4). Over time, the tiny clusters may stick to other clusters and so on, gradually increasing the average cluster size in the solution and resulting in large colloidal aggregates (Figure 6.8). The theory of cluster growth and aggregation is a rich area of physics, applicable to many soft matter systems, in particular colloidal particles suspended in a fluid. In this book, we will think simply about the process in terms of how spherical particles aggregate with time given an attractive potential between them.

The structure of a colloidal aggregate depends not only on the nature of the individual particles, but also strongly on the kinetics of the aggregation process (association and dissociation rates, for example). These rates will depend on the strength of the attractive forces holding the particles

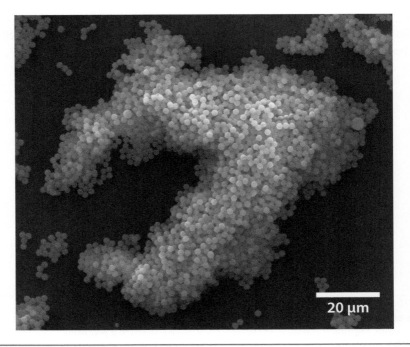

FIGURE 6.8   This scanning electron microscopy image shows a colloidal aggregate formed from 2-μm polystyrene beads. The beads are packed in an apparently disordered arrangement within the cluster.

together when they stick. If the forces between particles are weak, then particles may encounter each other on their random walk, but not always stick together, or they may have a high dissociation rate due to temperature-dependent Brownian fluctuations. Particles with a strong attractive potential will tend to stick permanently if they meet in solution. To see how these processes result in different aggregate structures, we can look at two limiting cases: reaction-limited aggregation (RLA) and diffusion-limited aggregation (DLA).

1. *Reaction limited aggregation.* In the reaction-limited regime, the kinetics of the aggregation process are most important in determining particle cluster structure. When two particles encounter each other in solution, there is some finite probability that they will stick to each other, but there is also a significant dissociation probability (or chance that they will not permanently stick). A simple method for visualizing the growth of an aggregate is to form a cluster by ballistic aggregation: dispersed particles are randomly added to the outside of the cluster, and the cluster will grow in size (Figure 6.9). As a cluster grows in size, vacancies inside the volume tend to form, and the growing aggregate will take on a fractal structure.[3] The interior structure of the cluster is dependent on the probability of a particle

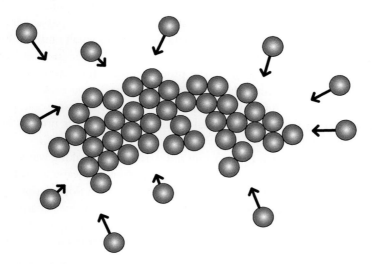

FIGURE 6.9 Ballistic aggregation is often used as a model system for colloidal cluster growth. The cluster is randomly bombarded by successive new particles that stick and gradually increase the size.

dissociating from another after coming close, or the reaction rate. Reaction-limited aggregates are fairly dense because particles are able to fill in some of the gaps in the growing structure; particles do not remain permanently stuck to every other particle they touch, so they can diffuse around inside the cluster.

2. *Diffusion-limited aggregation.* In a diffusion-limited regime, whenever two colloidal particles come close to each other they stick irreversibly, and the dissociation rate is zero. In this case, the rate of cluster aggregation is determined by the mean free time of the particles between interparticle collisions. This time is limited by the diffusion rate. DLA results in a more open cluster structure. As particles are added to the outside of the cluster, it becomes difficult for them to fill in any interior gaps without encountering another particle and becoming trapped. An analysis of diffusion-limited structures results in a fairly low fractal dimension, $D = 1.71$, indicating the high fraction of free space inside the aggregate.

Both of the models described can be used to represent the growth of a single particle cluster as single additional particles are added to the outside of the structure (Figure 6.9), and both processes result in a disordered fractal aggregate. In real solutions, we should think more about a *cluster–cluster aggregation* process in either the RLA or DLA regime as the mechanism for aggregation (Figure 6.10). Imagine a dispersed colloidal suspension with an attractive potential between particles at time zero. As time passes, first the particles will stick to their nearest neighbors to form small clusters. These small clusters will then diffuse around the solution until they encounter other small clusters and stick together. The process repeats until we end up with one large cluster spanning the material. Theoretically, we expect a fractal dimension of $D = 1.78$ for diffusion-limited cluster–cluster aggregation, and there have been many experimental demonstrations of this process in the scientific literature.

Induced *flocculation* is important industrially for drinking water purification and wastewater treatment. Suspended particles in water are often too tiny to be filtered out directly, so to solve this problem flocculation agents can be added. Flocculation agents often contain multivalent charged ions (alum, ferric sulfate, or aluminum chlorohydrate, for example) or charged polymers, and they induce aggregation of suspended particles into *flocs* by electrostatic screening.

A *floc* is a reversible aggregation of dispersed particles and can be easily filtered out. An alternative method for inducing flocculation involves adding polymers to the solution. This process works by taking advantage of the depletion forces discussed previously in this chapter.

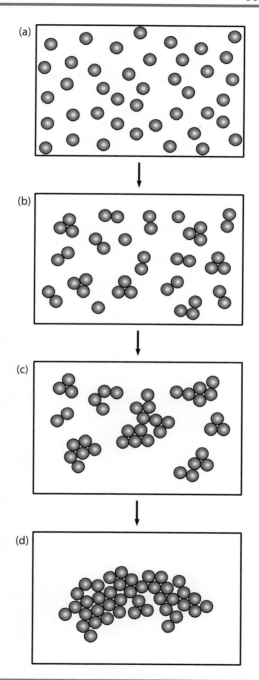

FIGURE 6.10   (a) Cluster growth by cluster–cluster aggregation. (b) Dispersed particles stick to each other, resulting in small clusters in solution. (c) These clusters then stick to other clusters in turn until (d) large aggregates are formed.

6.7 COLLOIDAL CRYSTALS

Colloidal particles in high concentrations can be thought of as model systems for crystalline materials. As they are pushed closer and closer together, particles will arrive at some optimal packing arrangement and form what is known as a *colloidal crystal*.[4] For a crystalline arrangement of particles with long-range order to form, the particles must be of uniform size (i.e., they have a low dispersity). In Figure 6.11, a scanning electron microscopic (SEM) image of densely packed polystyrene particles can be seen. These 2-μm particles are relatively uniform in size, and if you study the image closely, you can see domains of hexagonal and square packing between particles.

Just like the lattice structures in atomic crystals, colloidal particles can form equilibrium crystal structures by minimizing their interparticle potentials. Opals are naturally occurring colloidal crystals. These semi-precious stones consist of a close-packed structure of silica spheres and water, and

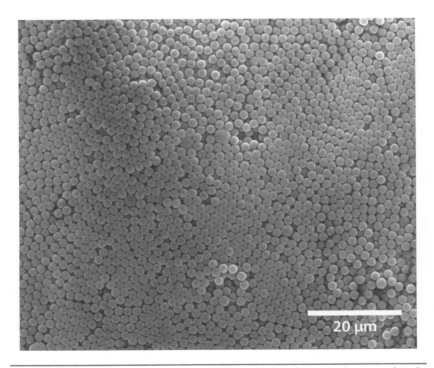

FIGURE 6.11   A scanning electron microscopic image of 2-μm polystyrene beads densely packed into crystalline-like domains formed by sedimentation.

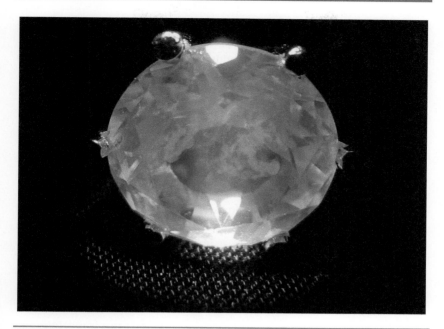

FIGURE 6.12 Striking reflection colors can be seen in this white opal, a naturally occurring colloidal crystal.

their often-colorful and iridescent properties are produced by Bragg reflection of light from the colloidal crystal structure (Figure 6.12).

For spherical particles, the most optimal packing structures that can form are the close-packed hexagonal arrangement or the face-centered cubic arrangement (both with 74% of space filled), although in practice other crystal structures can be formed. Colloidal crystals are potentially useful in the construction of optical devices known as *photonic crystals*.[5] In a photonic crystal, there is an optical band gap (i.e., a range of optical wavelengths that will not propagate through the material), and the colloidal length scale is perfect for designing devices to manipulate the transmission of visible light in a controlled fashion. The idea of an optical band gap is similar to the concept of a *phonon* band gap for lattice vibrations in crystalline materials or the reflection band of the cholesteric liquid crystal. In a colloidal crystal, a photonic (optical) band gap arises for photons with a wavelength comparable to the lattice spacing of the colloidal crystal structure. Bragg reflection prevents certain wavelengths of light from propagating. Such crystals can then be designed to act as waveguides by introducing line defects into the crystal structure. Colloidal crystals formed from silica spheres are particularly promising for future technologies, and by templating these close-packed colloidal structures one can form an inverse opal, a three-dimensional array of microcavities.[6]

6.8   GRANULAR MATERIALS

Any material that is composed of dry particles can be classified as a granular material. Examples from everyday life are numerous and include sand, sugar, coffee, and even M&Ms. When these materials are heaped up in piles or poured from one container to another, they exhibit flow, but do not behave exactly like a liquid. Understanding phenomena such as avalanches and jamming (e.g., particles spontaneously clog the mouth of an outlet when poured) has great importance for industrial applications. One of the classic problems in granular materials is the "sandpile," a simple heaping of sand or other granular material. What determines the stability of the pile and the onset of surface avalanches? Granular materials are not quite soft materials by our definitions at the beginning of Chapter 1, primarily because temperature is not important and therefore $k_B T$ is irrelevant to the material structure. The behavior of the granules in a sandpile is not governed by thermodynamics, instead particle cohesion and boundary conditions are important. Flows can occur when the sides of the pile are above a certain angle (the angle of repose), but otherwise the pile behaves as a solid. All the same, granular materials do exist as solid particles in a fluid phase (air), so they can be considered colloids.

To the casual observer, a granular material, such as the pile of salt in Figure 6.13, appears to be completely disordered and homogeneous. There is no obvious packing structure. However, if you look closely on the length scale of the constituent particles, the arrangement of these particles can be quite inhomogeneous, with particles forming locally connected "bridges" and other unstable, locally trapped structures. These structures are essentially short chains of particles supporting each other within the pile, similar to a keystone bridge; friction prevents the particles from slipping past each other, and force can be transmitted along these chains.

Jamming is a phenomenon that occurs when a granular material is poured through a narrow aperture. Initially, the granular material appears to have fluid-like properties, pouring smoothly, but then suddenly the outlet can clog and jam. This phenomenon is a serious problem for industrial processes that involve moving large amounts of granular material. A good reference to learn more about this field is the interesting review article "Granular solids, liquids, and gases" by Jaeger et al.[7]

FIGURE 6.13   Grains of crystalline materials such as salt or sugar are classic examples of granular materials. The material can become unpredictably jammed when flowing through a narrow aperture like the holes in this saltshaker.

6.9   FOAMS

A foam is an extremely light material composed of pockets of a gas phase dispersed in either a liquid phase (liquid foams like soap suds or the head on your glass of beer) or a gas phase dispersed in a solid phase (solid foams, i.e., your foam pillow or sofa cushions). It is interesting to think about foams (particularly liquid foams) as colloidal systems. The bubbles (gas phase "particles" suspended in a liquid continuous phase) move around in the fluid by Brownian motion and convection, they can also interact with each other by a variety of mechanisms (attractive and repulsive), much like the solid particles in a colloidal suspension. Foams represent a cross-disciplinary application of soft matter science, and so this section could equally belong in our chapter on surfactants because most foams include a surfactant of some kind to stabilize the bubbles and prevent the material from collapsing in on itself (Figure 6.14).

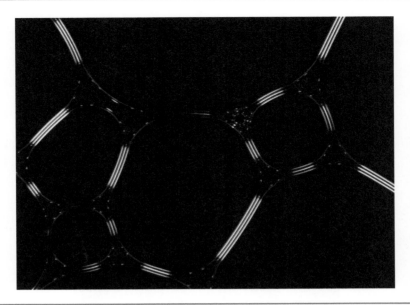

FIGURE 6.14   A soap foam formed from a simple mixture of ionic surfactant and water, imaged using polarized optical microscopy. The soapy films at the air–water interface are birefringent (see Chapter 3) and can be visualized using this technique.

The bubbles in liquid foams are often tightly packed and cannot slide past each other; this results in a more rigid structure than can be seen in the individual components. We can observe this quite easily in the washing up bowl. Both water and dish soap are liquids and therefore viscous, but when they combine to form a simple foam, solid-like elastic properties emerge. Under low shear, the foam can be formed into solid shapes (just like whipped cream or egg whites), but a high stress will produce liquid-like properties as the foam constantly restructures under shear. The viscoelastic properties of foam depend strongly on the "wetness" of the foam and bubble size. In a dry foam, the liquid fraction is very low, and the bubbles become close-packed polyhedra, compared to a high-water-content wet foam, the limit of which is a fluid-like bubbly liquid. For further reading, a detailed and simple description of the physics of foams can be found in the book *The Physics of Foams* by Weaire and Hutzler.[8]

6.9.1   WHY DO SOME LIQUIDS FOAM?

You will notice that it is fairly simple to introduce air bubbles into any liquid. Even plain water will form bubbles on the surface when shaken. After a few seconds at rest, however, the bubbles soon pop, and the liquid returns to a uniform state. The pressure difference across a gas–liquid interface can be described by the Laplace equation,

$$\Delta p = \frac{2\gamma}{r} \tag{6.9}$$

where γ is the surface tension of the liquid, and r is the local radius of curvature of the interface. This means that pressure from the gas phase either side of a liquid film in a foam will act to bring two bubbles closer together. Without a stabilizing mechanism, two adjacent bubbles will tend to rapidly coalesce, and the foam will not be stable. Foam stability, whereby the bubbles generated in a liquid remain for a long time, requires that we add an additional component to the system. For a foam to be stable, a mechanism must actively prevent bubble coalescence by providing a repulsive force between bubbles.

6.9.2 SOAP FOAMS

Soap foams are formed from a mixture of water and surfactant (for example, the common anionic surfactant, sodium lauryl sulfate; see Figure 4.1). As water drains from a foam and the material becomes mostly air, the fluid films between bubbles can become very thin. The hydrophilic surfactant head groups arrange themselves facing each other on either side of the water film with their tails oriented away from the water in a monolayer. In the case of sodium lauryl sulfate, the negatively charged head groups then repel each other. This provides a repulsive force between adjacent bubbles, preventing the liquid films from becoming so thin that the bubbles coalesce, collapsing the foam. For this reason, ionic surfactants are more effective at forming stable films than non-ionic ones.

You may have noticed that when you drop a bar of soap into your bubble bath all of the bubbles start to burst, and the foam collapses. This happens because surfactant from the soap bar (non-ionic sodium stearate, for example) rapidly begins to displace the bubble bath surfactant in the foam once entering the water. Because bar soap is not a particularly effective foaming agent, as soon as the two different surfactants are mixed, the original surfactant becomes diluted in the foam structure, and the foam begins to destabilize. We can estimate the thickness of the fluid films in a soapy shampoo foam, or even a single soap bubble, by observing the beautiful reflected colors. This phenomenon comes from thin-film interference; wavelengths comparable to the thickness of the film (~300–700 nm) constructively interfere to produce distinct reflection colors.

6.9.3   FOAM STABILITY

There are three main dynamic mechanisms governing the evolution and stability of a foam: *drainage, coarsening,* and *coalescence.* After bubble formation in a continuous liquid phase, gravity can play an important role in determining the foam structure. Shortly after bubble formation, the buoyant force on the gas pockets will cause bubbles to rise rapidly to the surface (provided the viscosity of the continuous phase is not too high). Bubbles push up against each other as they rise and can develop faceted shapes separated by thin liquid films. In a bubble bath, you may have observed that after some time the higher levels of the soapy foam floating on the water's surface are extremely light and almost seem dry. This effect occurs because excess fluid drains from the foam under the force of gravity. The draining process continues until the films between bubbles have thinned so much that other short-range mechanisms can come into play and stabilize the foam. Without an additional stabilization mechanism (for example, electrostatic repulsion), coalescence occurs rapidly at this point, resulting in foam collapse. Evaporation of the liquid phase also acts as a thinning mechanism, pushing the bubbles closer together.

Another phenomenon that occurs as the foam stabilizes is *coarsening* (often referred to as *Ostwald ripening*). This process is observed experimentally in many soft mixed systems. When pockets of one phase are present in another continuous phase (e.g., the gas bubbles in a liquid foam), gas molecules can diffuse between bubbles of higher pressure to those at lower pressure. This effect gradually increases the average bubble size in the system. If a bubble has a higher pressure than its neighbors, then this bubble will gradually shrink as gas diffuses through the thin connecting walls to the other adjacent, lower pressure bubbles. This ripening phenomenon is also seen in emulsions and aggregation processes. Ripening of the foam bubbles can be somewhat slowed by the presence of surfactant molecules at the bubble interface (Figure 6.15) or by adding additional ingredients designed to reduce gas diffusion in the continuous phase. Proteins can be a very effective foam-stabilizing material. If they are localized at the gas–liquid interface, the large molecules will provide a barrier to thinning and coalescence.

Another mechanism important in foam stability is the *Gibbs-Marangoni effect.* This phenomenon can prevent catastrophic thinning of the fluid films and subsequent bubble collapse. Consider two adjacent air bubbles in a foam, divided by a fluid film coated with surfactant molecules. As bubbles coarsen and grow, the bubble interface increases in area and the surfactant layer becomes stretched and thinned. Bubbles push closely on each other, creating flat facets and more curved vertices. This shape change means that the distribution of surfactant molecules coating the interface tends to decrease

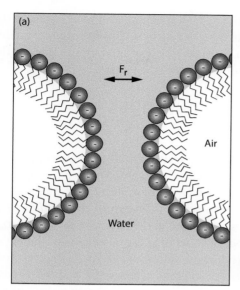 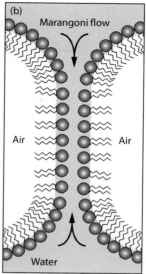

FIGURE 6.15 (a) Two bubbles coated with charged surfactant experience a repulsive force, F_r on close approach (b) Marangoni flow prevents coalescence in response to a surface tension gradient.

in density at areas of lower curvature (Figure 6.15). Surfactant thinning in these areas produces a surface tension gradient at the bubble/water interface, with the highest surface tension in the center of the bubble wall and lower surface tensions at the bubble vertices where the surfactant is not particularly thinned (recall that surfactants decrease surface tension at an interface). We learned in Chapter 4 that surface tension originates from a net force on surface molecules effectively pulling them into the fluid (cohesion), and that this force is greater for interfaces with larger surface tensions. Creating a surfactant gradient in the dividing bubble walls (from middle to corners) creates a surface tension gradient, thus providing conditions for Marangoni convection. The lower cohesive forces on water molecules below the denser surfactant layer produce a net flow of water (a mass transfer) from areas of low surface tension to areas of high surface tension. This effect can counteract the drainage of liquid from the bubble walls due to gravity and helps to explain the stability of foams formed with non-ionic surfactants, such as beer (the surfactants in that case are proteins).

The Gibbs-Marangoni effect can play a role wherever surface tension gradients are found. One classic example is the "tears of wine"—a high alcohol drink (e.g., wine or liquor) clings to the sides of a glass and drips down in a very different way to water. Dissolved alcohol reduces the surface tension of water and in a wine glass, the alcohol tends to evaporate fastest at the meniscus where the drink wets the glass surface. This generates a surface tension gradient and induces a mass transport of fluid up the sides of the glass—until gravity pulls it down (and we see the tears).

6.10   EXPERIMENTAL TECHNIQUES

6.10.1   LIGHT SCATTERING

Colloidal dispersions (unlike solutions) scatter visible light strongly. If you shine a laser pointer through a colloidal dispersion, you will see the path of the laser beam clearly as the incident light scatters from the (usually micron scale) particles. This phenomenon is known as the *Tyndall effect* and works because the particles suspended in the liquid are relatively large. You can observe the same effect by looking at car headlights in the fog or dust particles illuminated by a ray of sunlight. Colloidal particles can range from tens of nanometers to tens of microns in diameter, and the Tyndall effect is noticeable when the particles involved are of a size close to the wavelength of light. Light scattering can reveal useful information about the structure of a colloidal system, including particle packing structure and dynamics.

The use of light-scattering techniques to analyze solutions originates in the early work carried out by 19th century English scientist Michael Faraday, best known for his contributions to our understanding of electromagnetism. Faraday observed that solutions of colloidal gold appear red when illuminated with transmitted light, and that by adding sodium chloride the solution could be made to turn blue, until eventually the gold would precipitate out of solution.[9] Faraday determined that these surprising results derived from a change in the size of the particles suspended in the solution. The gold particles Faraday was working with were less than 100 nm in size and negatively charged; therefore, in a neutral solution they repelled each other and did not readily aggregate. Adding salt to the solution screened this electrostatic repulsion and induced the particles to aggregate—effectively increasing particle size. The light-scattering effect that Faraday observed was *Rayleigh scattering*. The Rayleigh scattering process is strongly dependent on wavelength, with a scattering intensity $I(\lambda)$ following the relation:

$$I \propto \frac{1}{\lambda^4} \tag{6.10}$$

This means that the intensity of scattered light is much higher for blue light than for red—the same effect that gives us a blue sky and remarkably colorful sunsets over Los Angeles and other polluted cities around the world.

When optical radiation of intensity I_0 is incident on a particle, if the particle is smaller than the incident wavelength λ, then an oscillating electric dipole is induced in the particle. This dipole radiates with the following intensity I distribution:

$$I = I_0 \frac{8\pi^2\alpha}{\lambda^4 r^2}\left(1+\cos^2\theta\right) \tag{6.11}$$

where α represents the polarizability of the particle, r is the distance from the particle at which the emitted intensity is measured, and θ is the angle between the incident and the scattered rays. For a uniformly dispersed dilute solution of small colloidal particles, the intensity of scattered light should follow this equation. For a solution of N identical particles, therefore, we can assume that the effect is additive, and the total radiated intensity I_N is equal to NI, where I is determined by the equation above.

Equation (6.11) describes diffuse scattering from a dilute solution of small particles, and the polarizability can be used to determine the average molecular weight of the small particles. However, much of the information we typically want to obtain from a scattering experiment is not contained in this equation. Light scattering becomes interesting when the size of the particle approaches the wavelength of the incident light. Scattering in this regime is known as *Mie scattering*, and in this case the intensity of scattered light can no longer be approximated using Rayleigh's formula. Mie scattering is not strongly wavelength-dependent, and the effects of this change in wavelength dependency can be seen in the white appearance of clouds and milk. These systems contain droplets of a size larger than the wavelength of light and scatter light of all wavelengths similarly. When observing a colloidal solution, the particle size can be often be approximated by considering scattering effects. If a fluid is illuminated by white light and has a cloudy appearance, this indicates that particle sizes suspended in the fluid are greater than the illuminating wavelength and probably micron-scale.

Both Rayleigh scattering and Mie scattering can be observed clearly in dilute solutions, but at higher colloid concentrations additional effects on the scattering profile come into play. Such effects derive from interparticle arrangements and are important in many industrially important colloidal solutions, such as paints and inks or different foodstuffs. High concentrations of particles will reduce the diffusion coefficient D of particles in the system and produce correlations between particles and collective behaviors. As a result of this self-organization, interference effects between scattered rays from different particles in the system will modify now the material scatters light.

6.10.1.1   Light-Scattering Experiments

Light scattering as an experimental technique can be used to investigate both the structure of a colloidal solution and more complex dynamic properties. Experiments can be divided into two different techniques, static light scattering and dynamic light scattering, although a combination of both techniques will often be used on a particular system. Raman scattering is also a light-scattering technique, but this has already been discussed in Chapter 5 in the context of polymers, so I won't describe it here. The wavelength of visible light is a suitable probe for particles ranging from about 50 nm up to tens of microns, so light-scattering techniques using visible wavelengths are ideal for characterizing colloidal solutions. The same techniques can also be applied to studies of the size of polymer molecules in solution, proteins, and aggregates of these molecules.

6.10.1.2   Static Light Scattering

A typical static light-scattering experiment is similar in concept to other particle- or photon-scattering techniques, such as electron, neutron, or x-ray scattering. Static light scattering is an elastic scattering process in which the frequency of the laser light used does not change. Solution samples are prepared, and a laser beam is oriented to shine through the sample (transmission geometry). The structure of the suspended particles can be investigated by analyzing the scattered light pattern as a function of scattering angle. Typical static light-scattering experiments probe particle or aggregate size in a dilute solution and can be used to study the distribution of particle sizes as a function of solution conditions or other parameters.

6.10.1.3   Dynamic Light Scattering

Dynamic light scattering (also known as photon correlation spectroscopy) is a measurement technique sensitive to the motions of particles in solution. As we learned previously in this chapter, all suspended particles in a colloidal solution are constantly subject to Brownian motion. Bombardment by solvent molecules leads to rotational, translational, and even more complicated conformational motions (in structured molecules such as polymer chains or proteins, for example). If we probe the solution with visible light, this constant particle motion will result in a time-varying fluctuation of the scattered intensity $I(\theta)$. Dynamic light scattering uses these fluctuations to extract information about the composition of the particle suspension from particle dynamics.

A dynamic light-scattering measurement is quite different from static light scattering. In the static experiment, the total scattered intensity at some angle is time averaged, and fluctuations in scattered intensity are not recorded. Dynamic light scattering uses the concept of a *correlation function* to recover information on the structure of the material by looking at how the scattering intensity fluctuates as a function of time.

Imagine a dilute solution of suspended particles. These particles will diffuse a mean-squared distance <R^2> in time t according to the diffusion equation:

$$\langle R^2 \rangle \sim Dt \tag{6.12}$$

where D is the standard diffusion constant. However, if t is small enough, particle positions in the solution will not have changed significantly from their initial state and therefore will be correlated with their original positions. After a long time, particle positions will bear no relation to the original positions (there will be no correlation between the current state and the initial state). The degree of correlation between any initial particle position state and a subsequent final state after some time interval Δt will decrease as Δt increases and can be modeled using an exponential decay function:

$$g(t) = A + Be^{-t/\tau} \tag{6.13}$$

Here, $g(t)$ is the correlation function, A and B are constants, and τ is a characteristic time for the exponential decay function, also known as the correlation time or the relaxation time. This characteristic time τ can alternatively be expressed as:

$$\tau = \frac{1}{q^2 D} \tag{6.14}$$

where q^{-1} represents a characteristic average distance traveled by a particle in time τ.

Another way of thinking about the correlation function is as a function that describes how the dynamic particle system evolves over time. Imagine taking a snapshot of all particles in the solution at some start time. A short time later, the particles will all have moved slightly in random directions, but their new positions will still be correlated somewhat with the original positions. Later, there will be very little correlation between the original positions as the particles move around randomly. τ is a time constant for the exponential decay of positional correlation.

We can relate this behavior to measurements from a scattering experiment. The scattered intensity from a solution detected at a fixed angle θ will fluctuate as a function of time because of particle motion, but if two intensity measurements are taken within a very short time interval, there should be some correlation between the two measurements. Essentially, if the time step is short enough, the particles will not have moved significantly from their initial positions. For a dilute monodisperse solution of similar particles, the time correlation function can be written as:

$$g(t) = A + Be^{-Dq^2t} \tag{6.15}$$

Figure 6.16 shows an example of a correlation function measured for a fluid containing monodisperse latex beads. For a *polydisperse* sample (i.e., particles with a size distribution), the analysis of scattering data is more complicated, but the molecular weight distribution can be determined by modeling the correlation function as the sum of the contributions from different discrete molecular weights.

We can obtain a direct measurement of the diffusion constant for the particles in solution by curve fitting the time correlation function and therefore obtain the particle radius R using the Stokes-Einstein formula:

$$R = \frac{kT}{6\pi\eta D} \tag{6.16}$$

where T is temperature, η is the shear viscosity of the solution, and D is the diffusion constant. This analysis assumes a spherical particle. Figure 6.16 shows an experimental example using this method can be used to measure the growth of colloidal aggregates over time (Figure 6.17).

For a more detailed description of light-scattering theory, there are many useful sources in the literature, such as Berne and Pecora's *Dynamic*

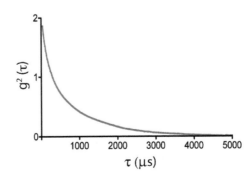

FIGURE 6.16 An example of a correlation function $g(\tau)$ collected in a dynamic light-scattering experiment on 100-nm monodisperse latex beads.

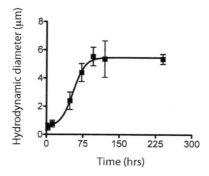

FIGURE 6.17 The results of a dynamic light-scattering experiment on colloidal organic carbon aggregation in Texas seawater. Aggregate size was measured as a function of time. (This seawater sample was collected from the Gulf of Mexico near Galveston Island, TX.) (Courtesy of Chi-Shuo Chen and Wei-Chun Chin.)

Light Scattering: With Applications to Chemistry, Biology and Physics[10] or the interesting chapter by Wu and Chu in *Experimental Methods in Polymer Science.*[11]

6.10.2 ZETA POTENTIAL AND THE ELECTRIC DOUBLE LAYER

When a charged colloid is placed in solution, the particle surface will influence the distribution of any ions in the surrounding fluid. Take, for example, a particle with an overall negative surface change. When this particle is placed in a solution containing both negative (anions) and positive charged ions (cations), these ions will respond electrostatically to the particle. The negatively charged particle surface will attract the positive ions from the solution (known as counterions) to its surface, forming a layer of charge around the colloid. These innermost charges are effectively bound to the surface in a layer known as the *Stern layer*. As we move away from the particle surface, this positively charged layer gradually decreases in density, forming a diffuse halo of counterions, the "edge" of which is called the slip plane. When the particle moves through the solution in response to the applied field, ions within the slip plane move with it. In turn, negative ions with a similar charge to the particle (known as co-ions) are repelled from the negative colloid and so increase in density away from the surface. This halo is known as the diffuse layer. The ionic arrangement described is known as the *electric double layer* and is depicted in Figure 6.18. This double layer effectively screens the particle's surface charge from the surrounding medium.

As a result of this charge distribution of counterions and co-ions, we define a surface potential for the particle. The *zeta potential* is the potential between the closely bound Stern layer and the rest of the suspending solution

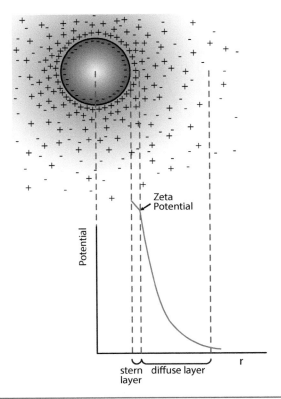

FIGURE 6.18   The electric double layer, comprised of the Stern layer and the diffuse layer, is defined by the charge distribution around a charged colloidal particle in an ionic solution. The electric potential around the particle is plotted here as a function of r, distance from the particle center.

(i.e., outside the diffuse layer). This modified potential defines how the particles tend to act in an ionic solution and can be measured by tracking their movement in an applied electric field. Charged particles move with a constant velocity under an applied external uniform electric field. The particles experience a Coulomb force and accelerate along the electric field lines, but they also experience drag forces from the suspending fluid. Because of these competing forces, the particles will reach a terminal velocity in the field and move with a constant velocity through the fluid. This process is called *electrophoresis*, and the velocity at which a particle will move depends on several factors: the size and zeta potential of the particle, the strength of the applied electric field, and the viscosity of the suspending fluid. Measurements are usually made in dilute aqueous solutions.

Although charge distributions on complex particles such as proteins or other anisotropic particle shapes can occur, most particles can be approximated as spherically symmetrical with a uniform charge distribution. In this case, we can use the Henry equation[12] to extract information about the particle:

$$\mu = \frac{2\varepsilon_r\varepsilon_0\zeta f\left(\lambda_D, a\right)}{3\eta} \tag{6.17}$$

Here, μ is defined as the electrophoretic mobility (particle velocity/applied electric field) of a particle of radius a. ζ is the zeta potential, and η is the viscosity of the suspending solution. $f(\lambda_D, a)$ is Henry's function and depends on λ_D, the Debye length (see Section 6.5.2). This variable represents the thickness of the electric double layer.

In everyday measurements, estimates for Henry's function can be used to simplify analysis of the zeta potential. If the Debye length is small compared to the particle radius, then $f(\lambda_D, a)$ tends toward unity, giving Smoluchowski's formula[13] for the particle mobility:

$$\mu = \frac{\varepsilon_0\varepsilon_r\zeta}{\eta} \tag{6.18}$$

In contrast, if the size of the electric double layer is large or the particles are small, then Henry's function tends to 2/3. This gives us the Huckel formula[14]:

$$\mu = \frac{2\varepsilon_0\varepsilon_r\zeta}{3\eta} \tag{6.19}$$

Zeta potential measurements are typically in the millivolt range, and a colloidal suspension with a zeta potential of less than 30 mV is usually considered stable, that is, the suspension will not aggregate spontaneously over a reasonable timescale. Of course, there are complications to this simple colloidal picture. Particles with complex charge distributions such as proteins or with an interesting surface topology (i.e., polymer brushes) may exhibit very different behavior compared to the ideal hard spherical colloid considered in this section. In these cases, a more detailed model for Henry's function is required.

The pH of the suspending solution is another important factor that can affect the zeta potential of a colloidal particle. High pH solutions tend to give the particles an overall more negative charge, and low pH solutions promote positive charge on the particles. At the *isoelectric point*, the zeta potential should be equal to zero, and the particle behaves as though it is uncharged. We can see the effects of this pH change in the kitchen. When milk, a colloidal suspension of protein particles, is mixed with a small amount of vinegar, the pH change induces aggregation in the solution, and curds are formed (Figure 6.19).

FIGURE 6.19   Milk is a colloidal suspension of oil droplets stabilized by globules of the protein casein that act as a surfactant. The suspended particles are larger than the wavelength range of visible light, so the white color of milk comes from Mie scattering from these oil-protein colloidal particles. Cheese is made by adding an acid to milk (e.g., lemon juice or vinegar). Casein is normally negatively charged in alkaline milk (the isoelectric point for casein is at 4.6), and electrostatic repulsive forces stabilize the colloidal dispersion. However, with the addition of the acid, the pH of the milk is reduced, and the casein becomes more neutrally charged. This change in the electrostatics of the system causes the oil-protein globules to flocculate and clump together, separating out from the aqueous whey into solid curds.

6.10.3   RHEOLOGY MEASUREMENTS

Rheological experiments are often used in colloidal science to character-ize the viscoelastic properties of suspensions. Colloidal systems are complex fluids, often with non-Newtonian behavior, and they can exhibit a variety of interesting phenomena. *Rheology* is the study of the viscoelastic properties of fluids, and by shearing a fluid in a controlled fashion, measurements are carried out to characterize the mechanical response of the material.

Many colloidal materials have unexpected properties at different shear rates. A thick solution of cornstarch and water can be so resistant to high shear rates that it behaves almost like a solid, but at slow shear rates would be easily characterized as fluid-like; this behavior is known as *shear thickening*. Alternatively, a material can be *shear thinning*; these materials flow easily at high shear rates. Canned whipped cream is an interesting example of a shear-thinning material (and also a foam). The foam will flow out of the aerosol can quickly under pressure, but when sheared very slowly behaves like an elastic solid. Shear-thinning materials can also be *thixotropic*. A thixotropic material shear thins but exhibits a time-dependent recovery in thickness once the shearing force is removed. Household paints and adhesives are good examples of thixotropic materials; they spread easily, but do not run or drip once deposited.

Rheology is an important aspect of colloidal science because many industrial products belong to this class of materials, and they all need to be transported along pipes or chutes, filled into containers, or poured and mixed at some point in their production. The behavior of a material under different flow conditions is therefore a parameter of practical importance because it determines how the material can be transferred and incorporated into products.

Some examples of typical rheological measurements in soft colloidal materials are

+ Quantifying viscoelastic properties by measuring bulk viscosity, elastic storage modulus G' and the viscous loss modulus G'' as a function of frequency
+ Measuring yield stress and material creep
+ Detecting structural transitions, for example, the onset of flocculation or the gelation point

These measurements of course can also apply to other soft systems such as polymers and surfactant phases.

6.10.3.1   Common Rheometer Designs

A *rheometer* is a piece of experimental equipment designed to measure the viscoelastic properties of a material, and in soft systems they can be used to characterize a variety of colloidal systems, including suspensions, emulsions, and pastes. Rheological instruments are also widely used in polymer science for polymer melts, rubbers, and solutions. The often-surprising flow

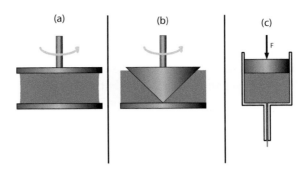

FIGURE 6.20   Schematic geometries of three different rheometer designs. The fluid under investigation in each case is shown in purple: (a) spinning disk rheometer; (b) cone-and-plate rheometer; and (c) capillary rheometer under applied force, F.

properties of many soft materials make rheometry an important industrial measurement because products must be transported, filled into containers, and dispensed. A variety of different instrumental designs for rheometers exist, and we review a few common examples here.

One of the simplest forms of rheometer is the capillary rheometer (Figure 6.20c). In this simple system, the material in question is driven through a well-defined capillary. The material enters the capillary at a fixed pressure, then the flow rate and pressure on exiting the exit aperture are measured. The resulting pressure drop ΔP across a capillary of radius r and length L and the measured bulk flow rate Q exiting the capillary are used to calculate the viscosity η of the solution using the following equation:

$$\eta = \frac{\Delta P \pi r^4}{8LQ} \tag{6.20}$$

Other common designs of rheometer use a shearing motion to characterize the material by rotating an object (a cup shape, plate, or cone, for example) in the material in a controlled fashion and monitoring the resulting torque on that object at a particular shear rate. In the spinning disk rheometer, a circular plate rotates on the surface of an enclosed sample, creating a constant shear stress (Figure 6.20a). For the cone-and-plate rheometer, the sample is prepared between a cone and a flat plate. The device senses the resistance to rotation of the cone within the material (Figure 6.20b).

Each of the rheological techniques described assumes that the material under investigation is homogeneous. These bulk measurements are useful for relatively large sample sizes and produce an average measurement.

In the case of other more delicate or locally inhomogeneous materials, different local rheological techniques can be employed in which tiny colloidal probe particles are dispersed throughout the material under study and their motions tracked and analyzed. These techniques are known as *microrheology*.

Microrheology is particularly appropriate for use in delicate materials that tend to break apart under small shear forces because it perturbs the material minimally. It can also be used with very small sample volumes, and is therefore particularly important for delicate biological samples, such as protein networks. Tiny tracker particles are dispersed throughout the material, and the motion of these particles is recorded over time. Using either particle-tracking software or scattering techniques such as dynamic light scattering, it is then possible to extract information on the local viscosity and elasticity of the material, presuming that the motion of the particles represents the properties of the material. The motion of the particles dispersed in the material can either be passively monitored (the random walk of the particle revealing viscosity using the Stokes-Einstein relation); alternatively, the particles can be actively displaced, and their response to the stimulus used to calculate various mechanical properties. You can learn more detail on the applications of microrheology to non-Newtonian materials from the excellent review article by Squires and Mason.[15]

QUESTIONS

Characteristics of Colloidal Systems

1. Investigate the relationship between surface area and volume in a colloidal system by starting with a 1-cm³ square block of material and gradually subdividing the block into smaller and smaller subunits. How does the ratio of surface area-to-volume scale with the particle size?
2. Explain why cities with a lot of suspended particles in the air have dramatic sunsets. If an intense bright blue sky appears to be a paler blue near the horizon, what can you deduce about the particles in the air?
3. What are the emulsifying agents in milk and mayonnaise, and why are these emulsions a milky white color? What are the characteristics of an emulsifier? Can you think of more examples in food science?

Colloidal Aggregation and Dispersion

4. Colloidal particles in suspension can be induced to aggregate by adding additional, smaller particles to the same solution. Explain this phenomenon.
5. Suggest two reasons why silt (suspended clay particles) deposits form at the mouth of a river delta.

6. Show that the terminal velocity for sedimentation of colloidal spherical particles in a *Newtonian fluid* can be given by the equation:

$$v = \frac{2R^2 \Delta \rho g}{9\eta},$$

where η is the viscosity of the fluid, ρ is the fluid density, and R is the radius of the particle. Calculate the terminal velocity for a polystyrene bead in water if the bead radius is 2 μm (the viscosity of water is 1.002×10^{-3} Pa·s, and the density of the bead is 1.05 g/cm³).

7. Large colloidal particles of radius A are mixed with smaller particles of radius B, resulting in a depletion attraction between the larger particles. Use geometrical arguments to derive a formula for the excluded volume of the larger particles, assuming they are spherical in shape.

8. Calculate the Debye screening length for a charged particle of radius 300 nm when dispersed in a 1 M NaCl aqueous solution at room temperature. How would your result change if $MgCl_2$ was used instead?

9. How does the Marangoni effect help to stabilize foams formed with a neutrally charged surfactant?

Experimental Techniques

10. How long will it take 200-nm particles added to one side of a 30-cm long horizontal pipe filled with water to diffuse to the other end in the absence of a flow? Make your estimate using the Stokes-Einstein relation.

11. A photonic crystal can be created by templating the crystalline packing of colloidal particles. For a close-packed, body-centered cubic arrangement of 100-nm spherical particles, what volume fraction is occupied by the particles? Predict the most intense reflected wavelength from this photonic crystal.

12. Dynamic light scattering can be used to determine particle size from the diffusion constant. In water, a silica particle diffuses on average, about 20 μm/s. Estimate the particle size assuming the particles are spherical.

REFERENCES

1. B.V. Derjaguin and L. Landau, Theory of the stability of strongly charged lyophobic sols and of the adhesion of strongly charged particles in solutions of electrolytes. *Acta Physiochim. URSS* 14, 633 (1941).
2. E.J.W. Verway and J.T.G. Overbeek, *Theory of the Stability of Lyophobic Colloids*. Amsterdam, the Netherlands: Elsevier (1948).

3. R. Jullien and R. Botet, *Aggregation and Fractal Aggregates*. London, UK: World Scientific (1987).
4. P.N. Pusey and W. van Megen, Phase behavior of concentrated suspensions of nearly hard colloidal spheres. *Nature* 320, 340–432 (1988).
5. E. Yablonovitch, Photonic band-gap structures. *J. Opt. Soc. Am. B Opt. Phys.* 10(2), 283–295 (1993).
6. A. Blanco, E. Chomski, S. Grabtchak et al., Large-scale synthesis of a silicon photonic crystal with a complete three-dimensional bandgap near 1.5 micrometres. *Nature* 405, 437–440 (2000).
7. H.M. Jaeger, S.R. Nagel, and R.P. Behringer, Granular solids, liquids, and gases. *Rev. Mod. Phys.* 68(4), 1259–1273 (1996).
8. D. Weaire and S. Hutzler, *The Physics of Foams*. Oxford, UK: Oxford University Press (1999).
9. M. Faraday, Experimental relations of gold (and other metals) to light. *Philos. Trans. R. Soc. Lond.* 147, 145 (1857).
10. B.J. Berne and R. Pecora, *Dynamic Light Scattering: With Applications to Chemistry, Biology and Physics*, 2nd ed. Mineola, NY: Dover (2000).
11. C. Wu and B. Chu. Light scattering. In *The Handbook of Polymer Sciences: Experimental Method in Polymer Science: Modern Methods in Polymer Research and Technology*. T. Tanaka, A. Grosberg, and M. Doi (Eds.). Boston, MA: Academic Press (2000), pp. 1–56.
12. D.C. Henry, The cataphoresis of suspended particles. Part I. The equation of cataphoresis. *Proc. R. Soc. Lond. Ser. A* 133, 106 (1931).
13. M. von Smoluchowski, Elektrische endosmose und Strömungsströme. In *Handbuch der Elektrizitat und des Magnetismus*. L. Greatz. (Ed.). Leipzig, Germany: Barth (1921), Vol. 2, p. 366.
14. E. Huckel. The cataphoresis of the sphere. *Phys.Z.* 25, 204 (1924).
15. T. Squires and G. Mason, Fluid mechanics of microrheology. *Annu. Rev. Fluid. Mech.* 42, 413–438 (2010).

FURTHER READING

T. Cosgrove, *Colloid Science: Principles, Methods and Applications*, 2nd ed. New York: Wiley-Blackwell (2010).

P.C. Hiemenz and R. Rajagopalan, *Principles of Colloid and Surface Chemistry (Undergraduate Chemistry Series)*, 3rd ed. Boca Raton, FL: CRC Press (1997).

R.J. Hunter, *Foundations of Colloid Science*. Oxford, UK: Oxford University Press (2001).

J.N. Israelachvili, *Intermolecular and Surface Forces*, 3rd ed. New York: Academic Press (2001).

T. Tanaka (Ed.), *Experimental Methods in Polymer Science: Modern Methods in Polymer Research and Technology* New York: Academic Press (1999).

T. Vicsek, *Fractal Growth Phenomena*, 2nd ed. London, UK: World Scientific (1999).

Soft Biological Materials

7.1 INTRODUCTION

Throughout this book, we have examined some basic classifications of soft materials, from liquid crystals and surfactants to colloids and polymers. At this point, you should be familiar with the characteristics of these materials. As you have seen, there is considerable crossover between these material classifications, but common themes unite the field. In this chapter, we will look at different soft materials that can be found in the living cell and explore new concepts in active matter using biological materials. Aside from our bones and teeth, the body is primarily composed of soft tissues, so we can naturally classify the building blocks of these tissues as soft materials, from the macroscopic level of fat and muscles, down to the microscopic structures inside individual cells.

The study of soft biomaterials is currently an active area of research in two different fields. *Biomechanics* is an engineering field that looks at how the body moves. This field concentrates on macroscopic materials and includes areas of research such as the development of prosthetics and other artificial biomimetic implants. *Biomolecular assembly* focuses on the physics and chemistry behind how particular biological molecules pack together and self-organize to form structures such as the cell membrane, protein filaments, and deoxyribonucleic acid (DNA). These two different fields deal with very different size scales, and in this chapter we focus on biomolecular assembly at micro- to nanoscale length scales. Biomolecular organization on these length scales down to the molecular level relates directly to the ideas of phase behavior and self-assembly that we have discussed throughout this book. The living cell is a vastly more complicated system than the simple equilibrium phases we discussed in the previous chapters. One should realize, however, that all of the structures in a living cell must arise from self-assembly processes of some kind. Molecular structures form or disperse as a result of intermolecular interactions and thermal fluctuations. Changes in the cell structure occur as a result of local concentrations of different molecules that activate or deactivate proteins. The contents of our cells are just a complex mixture of thousands of different kinds of molecules thermally fluctuating, associating, and dissociating with each other in a variety of different ways.

An important idea to introduce for this chapter, is that although living cells can be considered as made up of soft matter, this fact does not necessarily mean that those soft materials are stable phases in equilibrium. An equilibrium material or thermodynamic phase will always tend to adopt the most stable state (i.e., minimizing its free energy), and remain in that state unless perturbed by an external influence. Living cells do not behave this way, instead, they spontaneously adapt to their local environment and change internally all the time, creating local chemical gradients, transporting materials around their interior, and can even dramatically rearrange their contents to divide or move around.

Scientists often seek to apply the thermodynamics and mechanics of soft materials to biological systems, but this approach has some important limitations when we consider that the characteristic state of living systems is to respond to stimuli by consuming energy in the form of adenosine triphosphate (ATP), i.e., they do not come to, or maintain equilibrium. One of the most exciting developments in soft matter physics over the past few years has been the expansion of research into *active matter*, a concept which can be nicely applied to living materials.

In contrast to the "passive" materials that constitute the standard thermodynamic phases (gas, liquid, liquid crystal, etc.), active materials are not in equilibrium. Instead they consume energy and translate this energy into local motion. Various examples from biology can help to illustrate this idea. On the organismal scale, flocks of birds and insect swarms exhibit collective motion as the creatures self-propel, creating interesting internal dynamics within the group. On the cellular level, we can see similar collective behaviors where cells move together in clusters in response to a stimulus. Later in this chapter, we will look at an example of active matter on even smaller scales than this—collective motion of biopolymers. The unifying theme in all these active systems is that collections of subunits (birds, cells, biopolymers, etc.), take in energy locally, and then translate that energy into movement that can, in turn, produce large-scale dynamics. Internal motion throughout an active material can also result in the formation of emergent structures, such as moving topological defects (introduced in Chapter 3).

7.2 THE COMPOSITION OF THE CELL

If we want to discuss the biological building blocks of living organisms in the context of soft matter, we must first take a minute to consider a simplified version of the biological cell. In terms of building materials, what are cells? And, how does their makeup determine their properties and function? The cell is a highly complex environment in which the features we can observe under the microscope are continuously changing their form in response to various stimuli. However, certain structures are almost always present throughout the cell cycle, and they define the overall shape of the cell.

All cells are surrounded by a lipid bilayer, the cell membrane. They have a nucleus, containing the genetic material of the organism, and a cytoskeleton that provides shape, structure, and transport mechanisms and that facilitate cell division. Variants on this simplified structure make up each of the different tissues of the body, although there is a huge diversity in cellular form and function, depending on the specialization of that particular cell type (Figure 7.1).

There are many more fascinating elements to the cell than will be described in this chapter: mitochondria, ribosomes, protein ion channels, and the nuclear envelope, to name but a few. In this chapter, however, we restrict our discussion to the cell membrane, the cytoskeleton, and the nuclear

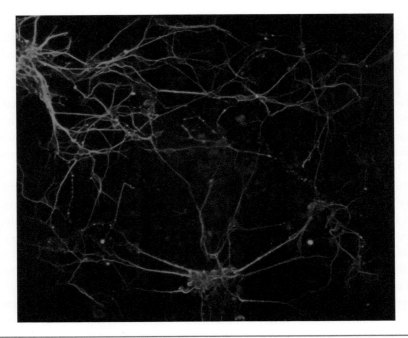

FIGURE 7.1 Neurons are highly specialized cells with an extended axon structure used to transmit electrical signals across distances. Here, we can see a confocal fluorescence microscope image of neurons stained with a green neuronal marker. The nuclear material is labeled in blue. (Courtesy of Wei-Chun Chin, University of California, Merced.)

material (e.g., DNA) as these systems are easily amenable to simple descriptions as soft materials. To learn more detail about cellular structures not covered here, some extra sources for further reading are suggested at the end of the chapter. After reading this book, I hope that you will start to think about molecular and cell biology in a different light and consider how cellular structures self-assemble as a result of intermolecular forces and thermodynamics.

7.3 THE CELL MEMBRANE

One of the most common soft materials in the body is fat, and fat is composed of surfactant-like molecules called lipids (Figure 7.2). Lipids in the body are not just present in the fatty deposits under the skin (with which we are often all too familiar)—they also form our cell membranes.

Scientific studies of the cell membrane go back to early cell biology, and in the 1920s, Gorter and Grendel[1] showed that the cell membrane was composed of a double layer of lipid molecules, or a *lipid bilayer*. In the lipid bilayer, molecular organization is similar to the lyotropic lamellar phase.

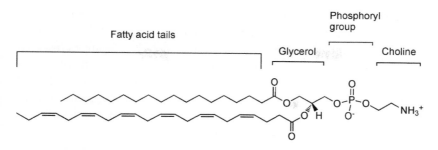

FIGURE 7.2   The structure of a phospholipid molecule with two different fatty acid chains: A saturated 18-carbon long oleic acid chain and the polyunsaturated DHA (docosahexanoic acid) chain.

Two fluid-like sheets of lipid molecules are arranged with the hydrophobic molecular "tails" pointing inward on both sides. Because lipids are surfactants, in aqueous solution they exhibit the bulk lyotropic phases described in Chapter 4. In the living cell, the membrane is just one bilayer enclosing the contents of the cell. This continuous bilayer envelope provides a selectively permeable membrane that wraps around the cell to form a barrier between the cellular machinery and the environment. As an analogy with surfactants, another way to think about the cell membrane is as a "bubble" under water, instead of the soap bubbles in air we are used to seeing. Most molecules cannot pass through the membrane by diffusion, including salts and proteins, so for molecular transport across the membrane, the cell employs a variety of different mechanisms. In addition to the simple bilayer structure, there are many dynamic interactions taking place in a real membrane all the time, including the action of ion channels, active and passive transport of molecules across the membrane, and interactions with the cytoskeleton to produce cell locomotion.

This idea of a cell membrane as a single bilayer bubble is of course highly simplified. The real cell membrane is not just a simple surfactant bilayer, but a complex structure composed of many different lipid molecules, cholesterol, and embedded proteins. However, despite the cell membrane's complicated composition, surfactant-like lipids largely define its physical properties.

In the following section, we consider a simplified view of the lipid bilayer as a model system for real cell membranes. We look at membranes composed of either one or a few lipid components to capture their underlying properties without worrying too much about the details of a real membrane. This strategy is known as using a *model membrane* and is employed by many experimental membrane researchers. An alternative methodology in membrane research is to extract the native membranes from biological cells. In this case, the complete membrane is extracted, including all of the different membrane proteins.

Such a system is most representative of the real thing, but results can be a challenge to interpret due to the complexity of the system.

The topic of biological membranes is particularly relevant to researchers interested in soft matter. The membrane is liquid crystalline (it's a two-dimensional [2D] fluid with short-range orientational order) with both lyotropic and thermotropic properties. In addition, the behavior of membranes modeled as continuous fluctuating elastic sheets as we described in Chapter 4 for surfactants is also relevant here.

7.3.1 Lipid Phase Behavior

In Chapter 4, we introduced surfactants and discussed how molecules with hydrophobic and hydrophilic sections will self-assemble into lyotropic phases in aqueous solution. The presence (or absence) of different surfactant phases (micellar, lamellar, etc.) depends on temperature, molecular shape, and concentration in water. In the discussions that follow, we focus on the phase behavior of a simplified model membrane containing only lipids, i.e., a model membrane. Figure 7.3 shows molecular structures for some common lipid molecules found in the cell membrane. For a detailed

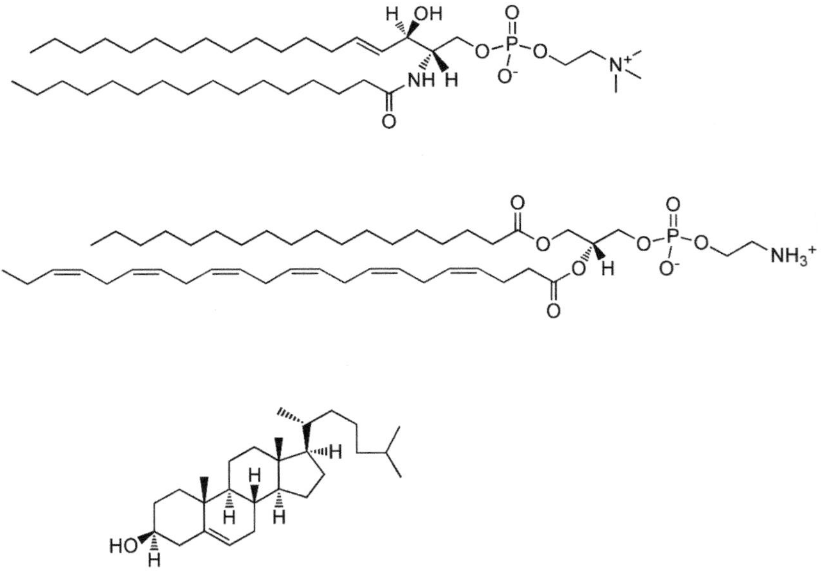

FIGURE 7.3 Molecular structures for some examples of common lipids found in the human body: top, sphingomyelin, [(2S,3R,4E)-2-acylaminooctadec-4-ene-3-hydroxy-1-phosphocholine]; middle, a phospholipid with one polyunsaturated chain (DHA); and bottom, cholesterol, a sterol also classified as a lipid.

description of the cell membrane, I recommend *The Structure of Biological Membranes* for further reading.[2]

In aqueous solution, lipids behave as *lyotropic liquid crystals*, forming a variety of different phases. We can observe the micellar, lamellar, and hexagonal lyotropic phases as a function of lipid concentration, although micellar phases are less likely to occur in a system composed of lipids with two flexible chains (due to their approximate cylindrical shape; see Section 4.6). Alternatively, lipids at low concentrations can be encouraged to form bilayer shells, these are known as either multilamellar or unilamellar vesicles. These lipid phases correspond directly to the surfactant phases that we discovered in Chapter 4. Lipid molecules are composed of a hydrophilic "head group" and typically two fatty acid "tails." The tails are hydrocarbon chains and may vary in "flexibility" at a given temperature depending on their degree of unsaturation (tails with unsaturated bonds are able to switch between different conformations at that bond, significantly decreasing the packing order of the membrane if the temperature is high enough to permit a significant number of conformational changes). The likelihood of a particular lyotropic phase occurring at a specific lipid concentration also depends on the molecular details of the lipid molecule, including average shape and charge. Molecules with an on-average cylindrical shape pack together well to form a flat bilayer and will tend to diffuse to low-curvature regions of the membrane, whereas a "cone-shaped" or "inverted-cone-shaped" molecule should pack more efficiently in a membrane with some curvature. For example, in Figure 7.4 we can see lipids arranged into the flat lipid bilayers that comprise the lamellar phase and a cross section of the lipids in a hexagonal phase.

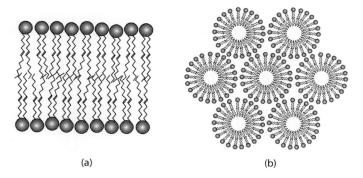

(a)   (b)

FIGURE 7.4   Cartoon models depicting the organization of lipids in cross section in aqueous solution for (a) a lipid bilayer and (b) the hexagonal phase. Individual lipids are shown in a simplified form here as a hydrophilic head group with two hydrophobic "tails."

In addition to their lyotropic properties, lipids can also behave like *thermotropic liquid crystals* in a similar fashion to the rod-like liquid crystal molecules described in Chapter 3. In aqueous solution, the primary organization of the lipid phase is lyotropic; however, the internal structural organization of the bilayer can vary as a function of temperature. In thinking about these thermotropic lipid phases, it is helpful to consider the lipid to be in the lamellar phase (i.e., a stack of bilayers), then to vary the temperature of this phase and look at the effects of temperature on molecular ordering within the bilayer. Lyotropic organization itself is also dependent on temperature, but more weakly so.

We can distinguish between the different thermotropic lipid phases by looking at the molecular arrangement in the plane of the membrane. For example, this could be the head group ordering, or chain packing in the hydrophobic core of the bilayer. The lowest-temperature thermotropic lipid phases are known as the subgel phases (L_c). These phases are pseudocrystalline, with solid-like properties. The molecules in the bilayer are extremely restricted in both rotation and translational diffusion. In the plane of the bilayer, the head groups have long-range order and a hexagonal packing arrangement. Lipid tails may also be tilted with respect to the layer normal. On heating the subgel phase, the next thermotropic phase to occur is the $L_{\beta'}$ phase, also known as the "gel" phase (and sometimes confusingly the "solid phase"). The gel phase is not typically observed in living cells although may have some biological importance. In fact, many of the lipids found in biological membranes would exhibit this phase at body temperature if they were in their pure state (i.e., not mixed with all the other lipids in the membrane). The gel phase has long-range orthorhombic in-plane ordering of the head groups with some degree of rotational freedom, although lateral diffusion is restricted. For many lipids, the molecular tails are significantly tilted in this phase, as shown in Figure 7.5. Above the gel phase in temperature, some lipid materials exhibit the unusual *ripple phase* $P_{\beta'}$. This phase is similar to the gel

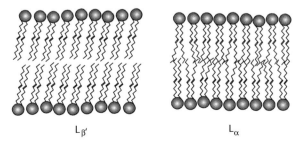

$L_{\beta'}$ L_{α}

FIGURE 7.5   Cartoon model demonstrating the differences in chain packing between lipid bilayers in the gel phase L_α (left) and the liquid crystalline phase $L_{\beta'}$ (right). Cross sections only are shown.

phase in viscosity, with an out-of-plane "ripple" in the membrane sheet, giving this phase its name.

At higher temperatures, the membrane will form the *liquid crystalline phase* (L_α), in which the lipid molecules are considered to be in a "fluid-like" state in the plane of the bilayer and diffuse freely. Lipid head groups take up a larger area in the L_α phase, and with short-range in-plane hexagonal lipid packing. In a bulk lamellar sample, this phase is the lyotropic equivalent of the smectic A (SmA) phase (Figure 7.5).

We can define a melting temperature, T_m, in a lipid system as the transition point from the gel phase (or from the ripple phase if present) to the liquid crystalline L_α phase. For biologically relevant mixed-lipid systems, this transition will occur below room temperature. Figure 7.6 shows an example of calorimetric (DSC—also see Chapter 3, Figure 3.32) data for the gel to fluid membrane phase transition ($L_{\beta'}$ to L_α). At 41°C, there is a distinct peak in the heat flow curve indicating a first-order phase transition with associated latent heat. This peak indicates the temperature (T_m) of the gel to fluid phase transition for the lipid sphingomyelin.

The liquid crystalline L_α phase is most relevant to membranes in living systems and represents a good description for the cell membrane, even though real membrane nanoscale structure may be much more complex. In the following section, we explore some of the interesting membrane phase behavior that has been seen in mixed-lipid membranes and its relevance to real biological systems.

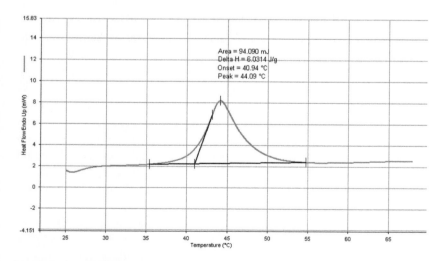

FIGURE 7.6 Differential scanning calorimetric (DSC) data for the saturated lipid sphingomyelin showing the phase transition from liquid crystal to gel phase measured with a heating rate of 20°/minute.

7.3.2   Lipid Domains and the Raft Hypothesis

The study of the cell membrane is an old area of research, going back to the origins of cell biology; however, there have been some revolutionary advances relatively recently in how we understand the function and structure of living membranes. So much research interest has been generated on this topic recently that it is worthwhile discussing the subject here.

Until the 1990s, the most widely accepted model for the cell membrane was the *fluid mosaic model*.[3] In this model, lipids and embedded membrane proteins were described as randomly distributed, with no particular in-plane distribution. This model is now considered to be overly simplistic, and current thinking is that biological membranes have a complicated in-plane organization. It is controversial whether or not the phase behavior of the constituent membrane lipids can play an important role in membrane function and protein organization, therefore the subject has generated a lot of research interest.

This new hypothesis for the structure of the cell membrane is informed by the phase behavior of mixed-lipid bilayers; phase separation of lipids might influence the organization of membrane proteins and ultimately their function. Membrane proteins could spatially cluster as a result of favorable interactions with lipid patches in the membrane of differing composition to the surrounding regions. This is known as the *raft hypothesis,* and these lipid domains, often referred to as "lipid rafts" or "lipid domains" are postulated to take the form of nanoscale patches in the membrane. The raft hypothesis is an attractive idea as it provides a simplified model for how protein organization may be controlled by local changes is lipid membrane composition.

The idea of spontaneous membrane phase separation being linked to cellular function was first introduced as recently as the late 1990s. In their influential article, Simons and Ikonen[4] described how certain mixtures of lipids can form in-plane bilayer domains. Phase separation is common in mixed systems, where the constituent molecules may have different affinities for each other. In a phase-separated system, two different phases can simultaneously occur at the same temperature in a material, usually with differing molecular compositions. Consider the simple example of oil and water (also Chapter 4, Figure 4.19, for example). Given enough thermal energy, they will mix; however, below a certain temperature, total phase separation will occur.

A lipid mixture designed to model the composition of the cell membrane will typically consist of some lipids with saturated chains (such as the sphingomyelin molecule shown in Figure 7.3), some lipids with unsaturated chains, and cholesterol, although the composition of a real cell membrane is much more complicated. Membrane phase separation in model systems can

FIGURE 7.7   Phase separation in a lipid bilayer. This giant vesicle is composed of a single bilayer formed from three different lipids. Phase separation into two different lipid phases is visualized using fluorescence labels. The red phase is the l_d phase, and we use a dye that does not partition into the l_o phase. Both phases can be visualized by using a different dye that partitions into both phases.

be observed using fluorescence microscopy. Below the *demixing* temperature (often denoted T_c), different coexisting phases can be visualized using suitable fluorescent labels (Figure 7.7).

The compositional phase diagram for lipid mixtures can be complicated, with regions of two-phase and three-phase coexistence. Most notably, a new fluid-like phase, liquid ordered phase (l_o), has been identified in these mixtures, with increased chain ordering in comparison to the L_α phase (also known as the liquid disordered [l_d] phase). Both l_o and l_d phases can occur in the same membrane in certain mixtures containing cholesterol. This liquid–liquid phase separation results from a selective partitioning of cholesterol in the mixture into domains with a high content of a higher T_m (typically saturated) lipid. Both of these fluid-like phases are liquid crystalline, with short-range in-plane ordering, although the l_o phase has an increased degree of chain ordering. Both of these phases are easily distinguishable from the gel phase with its extremely low lateral diffusion rate and long-range molecular ordering. In Figure 7.8, the schematic shows a cross section of a short section of membrane with phase separation into the l_o and l_d phases.

The possible membrane domains in the cell often described as lipid rafts are proposed to consist of the more ordered (l_o) phase. Both cholesterol and sphingolipids are hypothesized to be essential for lipid raft formation, and it is in systems where these molecules are present that much

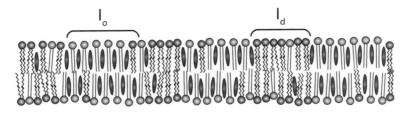

FIGURE 7.8   An example of lipid sorting in a three-component lipid bilayer. Domains of the l_o and l_d phases are depicted with lipids labeled as follows: green, high T_m; purple lipids, low T_m; blue lipids, cholesterol.

of the work on model membranes has been performed. Current interest in biological membrane rafts relates to their potential functional role in cell biology. It is currently thought that membrane domains in the cell are nanoscale in size, dynamically arising and dispersing as a function of local lipid composition. Additional theories examine how sorting of proteins via rafts may help to concentrate and localize proteins in the membrane, facilitating protein–protein interactions. The ordered lipid environment of the membrane domain may influence protein function, possibly by inducing a change in protein conformation. Techniques commonly used to study model membranes include fluorescence microscopy, nuclear magnetic resonance (NMR), atomic force microscopy (AFM), and x-ray diffraction (XRD).

7.3.3   MEMBRANE ELASTICITY AND CURVATURE IN BIOLOGICAL MEMBRANES

Biological membranes are composed of mixtures of different lipids, some with short chains, others with long flexible chains. The area per lipid head group can also vary depending on the specific molecule. The importance of molecular geometry for surfactant self-assembly was introduced in Section 4.5 when we discussed its role in the formation of different lyotropic phases. All surfactant monolayers can be described as having an intrinsic curvature c_0. This parameter describes the curvature of the membrane when free from external forces and is related to the average geometry of the lipid molecule (i.e., are they more cone-like, cylinder-like, etc.). Most biological membranes are composed of lipids which produce a membrane with an intrinsic curvature close to zero, promoting the formation of flat bilayers. Membrane lipids tend to be roughly cylindrical in shape (e.g., phospholipids with two hydrocarbon chains; see the sphingomyelin molecule in Figure 7.3). The intrinsic curvature of a biological membrane can have important consequences in terms of function, and by varying the lipid composition in a particular area of membrane, the cell may have a mechanism to control membrane shape.

If a membrane has no overall intrinsic curvature (i.e., its equilibrium shape is a flat plane), it costs energy to wrap that membrane around a cellular protrusion (or into a more complicated shape depending on what the cell is doing). Like a liquid crystal phase, undergoing a splay deformation, we can think about the lipid molecules splaying apart as a displacement from equilibrium. On large length scales (a radius of curvature of a few microns), the energy cost per lipid is not particularly significant, however, deformation around much smaller nanoscale structures such as cellular protrusions requires some significant molecular splay. To a first approximation, the cell membrane can be modeled simply as the elastic sheets we described in Section 4.7, describing resistance to bending in terms of elastic constants. The membrane is, however, in a fluid state (similar to a liquid crystal), and therefore the lipid molecules can diffuse around within the bilayer. Spontaneous localization of certain lipids which prefer a positive or negative intrinsic curvature (via diffusion) will help to reduce the elastic energy cost for a membrane to adopt a particular curved shape.

7.3.4   OTHER FATTY BIOLOGICAL MOLECULES

The lipids we have discussed so far in this chapter have been particularly relevant to the cell membrane, however lipids are also used to store energy in the form of body fat. Fat is stored in the body in the form of triglycerides and can also be present in blood plasma. These lipid molecules are insoluble in water and consist of a glycerol with three fatty acid tails. There are many different forms of triglyceride, and each of the fatty acid tails on the molecule are usually different, varying in length and degree of unsaturation.

Waxes are a variation of oil molecules and are the solid form of hydrocarbon alkane chains more than about 20 carbons long. Waxes are insoluble in water and usually amorphous plastic solids at room temperature. They are very hydrophobic and can be found in many plants and organisms acting as a waterproof coating. Some natural waxes include beeswax, lanolin, and carnauba. Paraffin wax is typically derived from crude oil for commercial use, although many plant waxes are very similar in composition, a mixture of mostly straight-chain hydrocarbons.

Leaves are protected by a waxy coating called the cuticle. This cuticle is composed of two waxy layers, a polymer matrix infused with waxes topped with a second wax layer. This coating provides the leaf with strong hydrophobic properties, allowing water droplets to run off the surface easily (see Figures 7.9 and 2.5).

FIGURE 7.9 Waxes are found widely in nature and provide plants with a protective waterproof coating. They are also flammable hydrocarbons and can be used as fuel.

7.4 PROTEIN STRUCTURES AND ASSEMBLIES

Proteins are large biological molecules made up of different combinations of amino acids, the basic protein building blocks. In eukaryotes, there are 20 different amino acids, and these small molecules are linked together into a long chain to form the protein (see Appendix E for a list of all the amino acids). Most proteins consist of a large number of amino acids with total molecular weights in the tens or even hundreds of kilodaltons. For example, G-actin, a common structural protein in the eukaryotic cell, has a molecular weight of 42,000 Da.

You may not have considered proteins to be a soft material before, but if you think about all of the different structures proteins can form (e.g., long polymer-like filaments, flexible sheets) and their behavior in solutions (as colloids and emulsifiers), then it is clear that they should naturally find a home in this field. Proteins are polymers and the amino acids are the monomers.

Proteins form the functional building blocks of our cellular machinery and have a complex hierarchical structure with different charge distributions and hydrophobic/hydrophilic domains on their surfaces. This structure cannot simply be described in terms of the amino acid sequence; the way the amino acid chain folds is as important for function as the sequence. Protein molecules have specific shapes tailored to their functions, including how they bind to other biological molecules. Protein structure is therefore complex and needs to be described in terms of primary, secondary, and tertiary structure.

The primary structure of a protein is the amino acid sequence. This is basically a list of the amino acids in the chain in the correct order. Each amino acid is commonly referred to by an abbreviated name (lysine is LYS, cysteine is CYS, etc.); Appendix E includes the structure of all of the amino acids with their three-letter abbreviations for your reference. Each amino acid has the same structural basis of a carbon attached to a carboxyl group and an amine group. They differ in the third attached group. Primary structure represents the smallest length scale of structure in a protein molecule, but an analysis of the amino acid sequence alone will not reveal complete information on the overall structure of the protein because this chain must next be folded into a specific shape.

Secondary protein structure defines the arrangement of the amino acid sequence (or primary structure) into larger motifs. The two most common ways that the amino acid chains can fold to form the secondary structure are into a *beta sheet* or an *alpha helix*. These secondary structures are held in place by hydrogen bonding. Some protein chain sections are intrinsically disordered. This means that they don't tend to fold in any particular way. Instead they have no fixed three-dimensional structure, similar to the random polymer coils described in Chapter 5. Many proteins incorporate intrinsically disordered sections into their otherwise folded structure.

The tertiary structure of a protein represents the fully folded structure of the protein sequence, that is, how the alpha helices, beta sheets, and other sequence units fold to give the overall shape of the protein molecule. Exactly how a given amino acid sequence can come together to form a final tertiary structure is a fascinating topic and an extremely rich area of active research. The tertiary structure is stabilized by electrostatics and weak bonds (disulfide or hydrogen bonds) between adjacent amino acid chains. If you start with a certain amino acid chain and look at minimizing the free energy of the molecule in solution, there may be many local minima in the energy landscape. Remarkably, a protein is capable of refolding into the same final structure reproducibly after unfolding. A protein must be folded correctly to function; however, misfolding of proteins does occur. On top of the tertiary structure there can be one additional level of

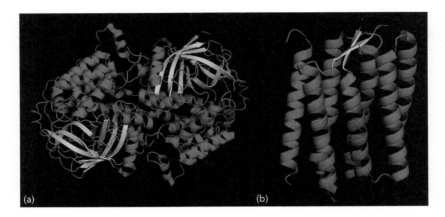

FIGURE 7.10 Ribbon diagram representations of protein structures for (a) bovine catalase (Protein Data Bank [PDB] entry 8CAT) and (b) bacteriorhodopsin (PDB entry 3WJK), a proton pump similar to the human visual pigment and found in bacterial cells. In these two examples of protein molecules, we can see the amino acid chain coiled and folded into different structures. Alpha-helical structures are shown in red and beta sheets in yellow. Diagrams such as this are commonly used to depict the complex hierarchical structure of proteins.

structure, the *quaternary structure*, in which different folded amino acid sequences of several protein subunits combine to form the overall protein molecule.

Figure 7.10 gives two examples of protein molecules that include prominent alpha helices and beta sheets. These ribbon diagrams are commonly used representations of complete protein molecules and represent the overall folded three-dimensional structure.

7.4.1 Protein Filaments

Protein filaments play many important roles in biological systems, and there are several examples associated with biological cells—the extracellular matrix, DNA, collagen, filamentous actin (F-actin), and microtubules, to name a few. These filaments can be thought of as biopolymers in that they have a repetitive long-chain structure comparable with the synthetic polymers we described in Chapter 5. There are important differences, however, between the freely jointed, flexible polymers we discussed previously and most protein biopolymers. First, most protein filaments are only semi-flexible (i.e., they are not freely jointed chains). In fact, from monomer to monomer, flexibility in a bulky biopolymer like F-actin or a microtubule can be very restricted, producing a filament that is relatively very stiff. This property significantly affects the range of molecular conformations possible when the chain is free in solution or introduced into confined spaces. Flexibility also strongly impacts how the filaments behave when cross-linked to each other or bound to other molecules. Second, the protein monomers in some biopolymers are very large, 1–2 orders of magnitude larger in size that the monomers of a polymer like polystyrene or polyethylene glycol. This difference has a big impact on diffusion rates, thermal fluctuations, and interactions with the surrounding solvent molecules.

Biopolymers are usually found as part of three-dimensional (3D) biological structures in combination with cross-linking proteins or other associated molecules. This ability to form macrostructures allows some protein filaments to act as structural materials for the cell. The most well-known example of structurally important biological filaments is the *cytoskeleton* of the eukaryotic cell. I have found that it is common for students in the physical sciences never to have encountered the cytoskeleton before, despite its critical importance to cell function and relevance to soft matter; therefore, we will spend a little time introducing the topic.

7.4.2 THE CYTOSKELETON

The cytoskeleton is a network of different biological filaments that crisscross the cell cytoplasm. These filaments form part of different cellular structures as they associate with each other and other related proteins. The different elements of the cytoskeleton have specialized roles in cell motility, transport, and division—without a cytoskeleton, cells would not be able to function. There are three main types of filament associated with the cytoskeleton: F-actin, microtubules, and intermediate filaments. Each of these filament types performs different functions within the cellular machinery and thus has very different mechanical properties.

The cytoskeleton could be considered a cross-linked polymer gel, but the real structure of the filaments that make up the cytoskeleton is much more complicated than our simple synthetic examples. In a living cell, the cytoskeleton is dynamic (not in equilibrium), and filaments continuously disassemble and reassemble in response to different stimuli. There are many different proteins involved with this reorganization and exact mechanisms are far too numerous to discuss here.

The protein actin is one of the most abundant proteins in the body and is probably best known for forming a major part of the muscle tissue. Actin also plays several critical roles on the cellular level as a component of the cytoskeleton. The basic form of the actin protein is the globular form, or G-actin, the structure of which is shown in Figure 7.11. Under the correct solution conditions, G-actin will assemble into F-actin. The globular G-actin subunits pack together helically to form a filament about 7 nm in width which can be up to about 20 μm in length.

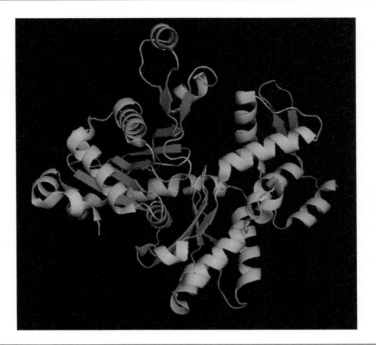

FIGURE 7.11 The molecular structure of G-actin, a globular protein with a molecular weight of 42 kDa. G-actin is the monomer for the biopolymer F-actin.

The detailed structure of the actin filament, however, is much more complicated than a typical polymer; because each individual subunit is comprised of a large G-actin protein (with a molecular weight of 42 kDa!), this is orders of magnitude larger than the simplest monomers we saw in Chapter 5 and as we can see in Figure 7.11, already very complex. The large size of F-actin is a great advantage to the observer as it makes imaging actin filaments much easier than synthetic polymers. Transmission electron microscopy (TEM) can be used to observe single actin filaments (Figure 7.12), and even though the physical size of the actin filament is much less than the optical resolution limit of light microscopes, the individual filaments may still be clearly observed with fluorescence labeling. This filamentous form of actin is a good example of a *semiflexible polymer*, which we discuss in the following section.

The *microtubule* is a hollow tubular filament that assembles under the correct conditions from the protein tubulin and can be found fulfilling many different biological functions. Microtubules provide the cell with "tracks" along which molecular motors can transport vesicles containing cargo. They also facilitate chromosome rearrangement during mitosis and form

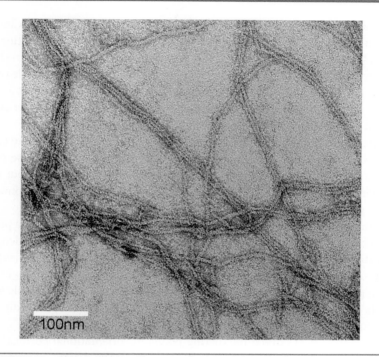

FIGURE 7.12 A transmission electron microscopic image of single F-actin filaments. The filaments are negatively stained with uranyl acetate.

the cilia or flagella of bacterial cells. Two forms of the tubulin protein, alpha and beta tubulin, form a dimer, and these dimers assemble into the protofilaments that form the microtubule. Other biological polymers include polysaccharides, glycogen, and collagen. DNA is arguably the most well known and most studied; we discuss this interesting molecule in a bit more detail in Section 7.4.4.

7.4.3 SEMI-FLEXIBILITY AND PERSISTENCE LENGTH

Many biopolymers can be classified as semiflexible filaments. To understand this concept, we can start by thinking about the limits of flexibility in a general long molecule. The polymers we discussed in Chapter 5 were long, freely jointed chains that tend to adopt a globular configuration in solution with a characteristic radius of gyration. The thermal environment constantly bombards a floppy chain with molecular collisions to produce strong shape fluctuations. In the case that a chain is very flexible (freely jointed), neighboring segments can adopt any orientation, and we could say that as we move from one segment to the next, the chain "forgets" its orientation, or that the orientation of the molecule does not "persist." The other limit of flexibility is a rigid rod, with a fixed, straight geometry. In this case the orientation of the first segment is the same as that of the last. The semi-flexible regime lies somewhere between these limits, where the characteristic length scale over which a filament loses its orientational information is more than just a few monomers and can be of the same order as the length of the filament or even longer. This length-scale is known as the *persistence length* (L_p).

Actin filaments are more flexible than microtubules. DNA is quite flexible, but still can be considered a semi-flexible biopolymer. For these biological semi-flexible rods, bending fluctuations are still important, but the filaments do not form the globular random walks we described for conventional polymers.

The persistence length, L_p, provides a measure of the stiffness of a polymer chain by looking at its thermal shape fluctuations. Persistence length is particularly useful in the case of semiflexible filaments where the length scale of curvature is significantly more than a single monomer segment spacing and may be comparable with filament length. For semi-flexible filaments that remain relatively straight, end-to-end distance and the radius of gyration are not particularly useful concepts.

In a system of curved fluctuating polymers, orientational correlations between chain segments decay exponentially with the distance separating them, and we can define persistence length by:

$$\left\langle \hat{s}_1 \cdot \hat{s}_2 \right\rangle = \left\langle \cos\theta \right\rangle = e^{-L/L_p} \tag{7.1}$$

where s_1 and s_2 are unit vectors representing bond orientations of two segments on the chain, separated by a contour length ΔL (Figure 7.13). θ is the angle between s_1 and s_2, and L_p is the persistence length of the filament.

As in all polymer systems, it is important to remember that we need to take a statistical approach when thinking about persistence length in biopolymers. Polymers fluctuate thermally and adopt random chain configurations

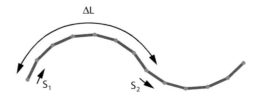

FIGURE 7.13 Schematic representing the definition of persistence length for a flexible filament: the average contour length (ΔL) over which orientational vectors, **s**, on the chain become uncorrelated.

that fluctuate over time. A snapshot of many similar filaments will reveal chains with a variety of different configurations, from quite bent to relatively straight. Persistence length provides us with a measure of the *average* contour length over which the chain loses orientational information, and thus provides a characteristic length that can be used to characterize the flexibility of the filament.

For example, at the same temperature, a population of floppy filaments will tend to exhibit much more curved configurations on average than those produced by stiffer filaments. Or, the average distance ΔL, you need to travel along the floppy filament before two points, defined by orientation vectors, s_1 and s_2 (Figure 7.13) are orthogonal, will be much shorter than that for a stiff filament. In this definition, the persistence length is a correlation length between two tangential vectors along the filament.

Molecules with a short persistence length (e.g., a conventional polymer, such as polyethylene) need very little energy to produce curvature over short contour lengths; the persistence length of a freely jointed chain at room temperature is close to the intermonomer distance (the segment length). In biopolymers, persistence lengths can be very long because the chains are much stiffer and therefore typically more extended in solution. Microtubules are the stiffest biopolymers in the cell with a room temperature persistence length of about 1 mm. F-actin has a persistence length of about 10 μm.

A more detailed mechanical analysis of filament fluctuations (you can read more on this subject in the excellent book by Boal, *Mechanics of the Cell*) gives us an alternate definition of persistence length related to the bending rigidity, B of the filament,

$$L_p = \frac{B}{k_B T} \tag{7.2}$$

In general, bending a uniform elastic rod costs energy, and the increase in potential energy, U associated with the bend can be calculated using a form of Hooke's law,

$$\Delta U = \frac{1}{2} \frac{BL}{R^2} \tag{7.3}$$

In this formula, R represents the radius of curvature for the arc of the curved rod, and L is the length of the rod. B is equal to YI, where Y is the Young's modulus of the rod and I is the moment of inertia ($1/4\pi r^4$ for our uniform elastic rod of radius r).

Stiffer filaments (higher bending rigidity, B) are expected to have a longer persistence length than more flexible ones. It takes energy to bend a filament, and in a thermally fluctuating system this energy comes from the environment. Higher temperatures will produce a greater incidence of filaments with greater bending fluctuations (and thus a shorter persistence length).

7.4.4   MICROTUBULES AND MOLECULAR MOTORS: AN EXAMPLE OF A BIOLOGICAL ACTIVE NEMATIC

At the beginning of this chapter we discussed the idea that our cells are not in equilibrium. A system in equilibrium reaches a steady state and then tends to stay in that state. Cells of course do not exhibit such behavior and constantly adapt and change their structure in response to stimuli. Active systems consume energy (for example, in the form of ATP—chemical energy) and somehow translate that energy into mechanical energy to produce internal dynamics and collective motion. Under this definition it is clear that living systems are not particularly well described as equilibrium matter - they are active matter. Active matter is very unlike a conventional equilibrium phase, which does not need any energy to maintain its thermodynamic state.

To form an active material, we first need a collection of subunits (for example, cells or biopolymers) and a mechanism for subunit propulsion. In our cells, there are tiny motor-like molecules (such as kinesin and myosin), which use ATP (a form of chemical energy) to generate a mechanical motion. Specifically, these motors are able to take steps along a biopolymer filament, such as F-actin or a microtubule. This stepping motion (converting chemical energy to mechanical energy) plays a role in some very important biological functions. For example, kinesin motors facilitate the transport of chemical cargos around the cell by "walking" along the microtubules in 8 nm steps. There are many other molecular "motors" in biology - tiny nanomachines that have the ability to perform mechanical actions. We are only just beginning to understand their potential uses in synthetic active materials. Scientists recently took advantage of this interesting mechanical mechanism[5] to produce a new form of liquid-crystal-like active matter. They formed a two dimensional, fluid-like phase in which densely packed microtubules slide relative to each other. The sliding motion is continuous and powered by the action of kinesin molecular motors. Figure 7.14 shows an example fluorescence microscopy image of the material. In this image the biopolymers (microtubules) form bundles that bend and break continually once in the nematic arrangement. Topological defects then spontaneously form as a result of the global sliding motions (see if you can identify +1/2 and −1/2 defects in the image - see Section 3.6.1). The defects move around each other and continually form and annihilate with each other producing a fascinating dynamical out-of-equilibrium state.

FIGURE 7.14   An example of biological active matter. This system is a two-dimensional nematic-like phase consisting of fluorescently labeled microtubules (bundles of which are seen clearly in the image) and kinesin molecular motors (not visible). The system exhibits continuous internal dynamics, including the spontaneous formation and annihilation of topological defects as long as the energy source (ATP) is available. (Image courtesy of Amanda Tan, University of California, Merced)

7.4.5   THE NUCLEIC ACIDS

DNA and riboxynucleic acid (RNA) are arguably the most important molecules for life and among the most fascinating. These long polymeric molecules are found in the cell nucleus, and their unique structures encode the genetic information of the organism. It would be remiss not to include a discussion of DNA in this chapter because so much recent work has gone into the uses of DNA in biotechnology and new biomaterials. As biological molecules that will self-assemble in solution into different structures, the nucleic acids can easily be considered "soft" in the same way that we consider F-actin and other structure-forming proteins to be soft materials. Both DNA and RNA are long, semi-flexible biopolymers, and their behavior in solution can be modeled as a simple polyelectrolye.

7.4.6   THE STRUCTURE OF THE NUCLEIC ACIDS

Both DNA and RNA are known as nucleic acids, and they have a complex structure designed for their specific biological function of information storage and replication. There are two basic elements to the DNA molecule, the

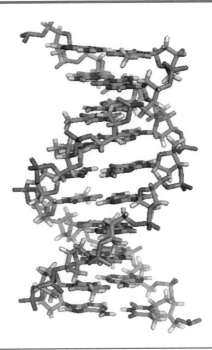

FIGURE 7.15 DNA is a double-stranded helical molecule composed of two intertwined sugar–phosphate backbone strands connected by the base pairs. Here, the elements are displayed by color; carbon is green, nitrogen is blue, oxygen is red, and sulfur is orange. (Adapted from Protein Data Bank entry 1 bna and rendered using Pymol.)

backbone and the base pairs. The backbone of the DNA molecule is composed of two helically twisted strands: the famous "double helix" (see Figure 7.15). Each of these strands is a chain of deoxyribose sugar segments linked by phosphodiester bonds. This is commonly known as the "sugar–phosphate backbone." Attached onto each of the sugar groups is one of four different bases: adenine (A), guanine (G), thymine (T), or cytosine (C). The order in which these bases occur along the chain encodes the genetic information for the organism. Each sugar–phosphate segment with attached base is known as a *nucleotide*.

The two twisted chains of the DNA molecule are held together by hydrogen bonds between the bases. Wherever an *A* nucleotide protrudes from one chain, a *G* will appear on the other chain, and the two are linked by a double hydrogen bond. Similarly, wherever a *T* occurs, it will be matched on the other chain by a *C* and linked with a triple hydrogen bond. This matching means that both nucleotide chains in the double helix carry the same encoded information, and this is important in the process of replicating the DNA code.

There have been many attempts in recent years to use the exact pairing properties of the DNA molecule to create engineered nanostructures, and now it is possible to design and synthesize short custom DNA molecules with a predefined sequence and build up tiny tunable arrays or patterns from the DNA building blocks.

7.5 EXPERIMENTAL TECHNIQUES

Biological molecules can come together to form a huge variety of soft materials with two-dimensional and three-dimensional structures ranging from the nanoscale to the macroscale. These can include structures found in the biological cell, or large-scale phases and assemblies based on biological extracted molecules. Many biological assemblies are just like the synthetic soft materials we discovered earlier in this book. Lipid membranes, protein filaments, etc. can be assembled in the lab from their constituent molecules and studied just like their synthetic counterparts (surfactant film and polymers). We can also take molecules extracted from biological systems and use them to assemble totally new structures and phases not found in nature! In this section, I have chosen a few examples of biological materials to illustrate the parallels between biological assemblies and other soft materials. I also introduce a variety of techniques often used in the preparation and characterization of soft biological matter.

7.5.1 STUDYING MEMBRANE BEHAVIOR

Biological membranes are delicate structures that assemble as a result of the amphiphilic properties of lipid molecules; in an aqueous environment, they will form lipid bilayers. Many laboratories around the work are interested in the intrinsic properties of biological membranes, such as their physical structure, transport mechanisms across the membrane, and membrane proteins. There is also much interest in using membranes (either biological or bioinspired) for engineering applications such as drug delivery and separation science. To investigate the properties of a lipid membrane, one must first decide how to prepare the membrane system, and there are several different geometries that are commonly used experimentally. Because single lipid membranes are so delicate, surface forces and hydration can have a significant effect on their structure. In the following sections, we take a look at some different membrane geometries used experimentally.

7.5.1.1 Lipid Vesicles

Lipid vesicles (also known as liposomes) are spherical membrane shells of lipid bilayer in solution. You could think of them as a bubble with water inside and out. Vesicles are stable with aqueous solutions on either side of the membrane provided the osmotic difference between the two solutions is not too large. Vesicles can be large or small, multilamellar or unilamellar, and provide us with a useful experimental tool. In biotechnology, liposomes are widely used to encapsulate specific molecules and can provide a mechanism for drug delivery with a targeted or slow release.

Giant unilamellar vesicles are often used to study phase separation phenomena in membranes, membrane stiffness, or porosity. These large unilamellar vesicles are single bilayer spheres with a diameter that can range from about 5 to 50 μm (Figures 7.7 and 7.16). Giant unilamellar vesicles can be observed with bright-field microscopy, although they are difficult to see because the single bilayer is only about 6 nm thick. Visualization can be greatly enhanced by the use of either phase contrast or DIC (differential interference contrast) microscopy or fluorescence microscopy if a fluorescent probe can be incorporated into the lipid bilayer.

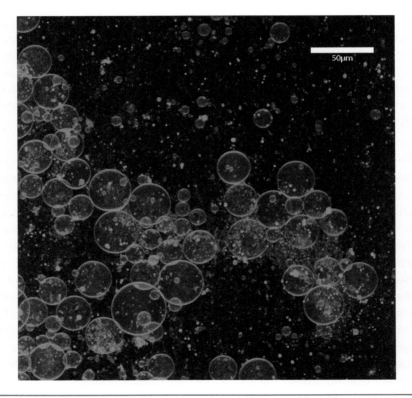

FIGURE 7.16 A fluorescence microscope image of giant unilamellar vesicles (GUVs) in water. These vesicles were prepared from the lipid DOPC (1,2-dioleoyl-sn-glycero-3-phosphocholine) and visualized by including a small proportion of fluorescent lipid in the bilayer (<0.05 mol%).

7.5.1.2 Imaging Membranes Using Atomic Force Microscopy

AFM is an experimental technique that uses an extremely sharp probe to measure the nanoscale structure of surfaces. Membranes can have an interesting structure on this length scale if there are lipid domains or transmembrane proteins present, for example, the lipid rafts discussed in Section 7.3.2 and associated proteins. AFM can be used to map the topology of a model lipid bilayer or extracted biological membrane if the membrane can be transferred to a very smooth surface.

The AFM produces an image by moving a very sharp tip across a surface; this can be the top surface of a lipid bilayer deposited on a substrate. The AFM tip, mounted on the end of a flexible cantilever, is deflected up or down as it scans across the surface of the sample, mapping the surface topology. As the tip is deflected, the position of a laser beam reflected from the upper surface of the cantilever is detected by the instrument. The deflection of the reflected laser beam is used to calculate the height profile of the surface and produce a topographical image (Figure 7.17). The AFM tip usually has a radius of curvature of just a few nanometers and can easily achieve an in-plane resolution in fluid (membranes must be kept under water at all times to preserve their hydrated bilayer structure) of about 1 nm. There are a variety of different modes in which an AFM can operate to map different characteristics of a surface; these include lateral force microscopy (LFM), sensitive to surface roughness and electrostatic mapping. The AFM tip can also be used to quantify the elastic response of the membrane in a particular area by looking at laser deflection as the probe is lowered and pushed onto the surface (force spectroscopy).

To collect surface data on native biological or model membranes using the AFM, a uniform lipid bilayer must be deposited on a substrate, and typically a single bilayer is preferred. The most common substrates are freshly cleaved mica, a fresh graphite surface (stripped with tape), or a cleaned silicon wafer. Freshly cleaved mica and graphite provide an atomically smooth substrate and therefore the best height resolution. Membranes deposited on a surface for this technique are known as *supported bilayers*. There is some question that a hard substrate may modify the properties of the membrane when compared to a free membrane, and that the molecules of the lower leaflet may not be as free to diffuse as they should be. As a remedy to this problem, membranes can also be formed on a polymer cushion to remove the effects of the underlying hard substrate. A polymer layer is first deposited onto the substrate, providing a separation between membrane and the hard surface; then the lipid bilayer is prepared on top of this polymer layer.[6]

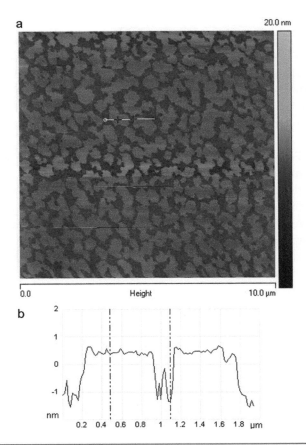

FIGURE 7.17   An atomic force microscope (AFM) image of a single lipid bilayer prepared on a mica surface from a mixture of the lipids DOPC (1,2-dioleoyl-sn-glycero-3-phosphocholine), DPPC (1,2-dipalmitoyl-sn-glycero-3-phosphocholine), and cholesterol. (a) A topology map of the membrane surface, with the liquid crystalline L_α phase/gel phase $L_{\beta'}$ coexistence clearly seen as light and dark regions on the image. (b) A plot of the height profile across the line marked on the image. Notice that the regions of the membrane in the gel phase are about 1 nm thicker than membrane in the liquid crystalline phase (see Figure 7.8).

7.5.2   Fluorescence Microscopy

Many molecules will absorb light at specific optical wavelengths. *Luminescence* is a phenomenon that occurs when an electron in a material is excited to a higher energy state by an incident photon. The subsequent relaxation process back to the ground state results in the emission of a photon. *Fluorescence* specifically comes from the excitation of an electron to an excited singlet state (in which the excited electron remains spin paired with one in the ground state). The decay process is fast, taking place on a nanosecond timescale. In comparison, the *phosphorescence* process involves excitation to a triplet state (the excited electron is also spin flipped), and relaxation is much slower.

In fluorescence microscopy (Figure 7.18), we take advantage of the energy change that occurs between the absorption and reemission of the incident

FIGURE 7.18 A simple upright fluorescence microscope can be used to image fluorescently labeled biological molecules, either *in vitro* or in cells. This microscope has a reflection configuration (emitted light is reflected back through the objective to the eyepiece), also known as epifluorescence.

photon in a particular molecule. The reemitted photon has a lower energy, E, and so a longer wavelength, λ, than the exciting photon since:

$$E = \frac{hc}{\lambda} \tag{7.4}$$

where h is Planck's constant, and c is the speed of light in a vacuum. We can take advantage of the wavelength shift between excitation and emission to image biological molecules that would otherwise be invisible to bright-field microscopy techniques or to highlight the location of certain biological structures (Figure 7.19). By using filters on the microscope, it is possible to only observe wavelengths emitted from the molecules of interest, providing an image representing their locations. Figure 7.20 demonstrates the difference in excitation and emission spectra for a common fluorescent probe. Fluorescence microscopy is one of the most common imaging techniques used for biological materials because of the ease with which a variety of fluorescent molecules (also called probes or labels) can be incorporated into the system.

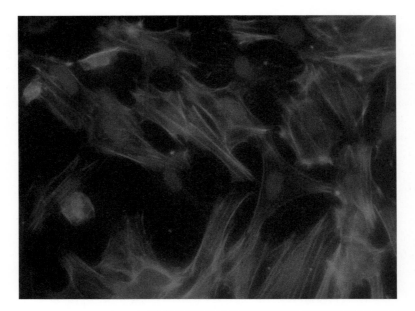

FIGURE 7.19   A fluorescence microscope image of stem cells. The F-actin cytoskeleton is labeled in green and the nuclear material in blue, indicating the position of the cell nucleus. (Courtesy of Chi-Shuo Chen and Wei-Chun Chin, University of California, Merced.)

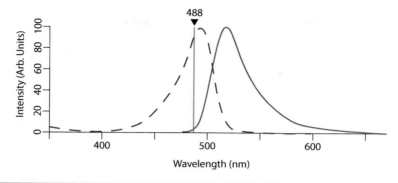

FIGURE 7.20 Excitation and emission spectra for the fluorophore Alexa Fluor 488. Notice how the excitation spectrum (blue dashed line) does not significantly overlap the emission spectrum (red solid line).

There are many different microscope configurations that can be used in fluorescence imaging, but they all operate on the same basic principle; a fluorescent molecule is introduced with the goal of highlighting a particular structural feature of the system. For example, in Figure 7.19 some cells have been labeled with fluorescent probes to highlight the nucleus and the cytoskeleton. This was achieved by introducing fluorescent molecules designed to migrate to those specific areas. The blue probe binds to the nuclear material highlighting the nucleus, and the green probe is bound to the actin filaments. As an imaging tool, fluorescence microscopy is commonly used for visualizing cellular components, but can be used to investigate the distribution of materials in any optically transparent system, such as *in vitro* protein gels, membranes, or colloidal systems.

There are thousands of known fluorescent probes, including both intrinsic (naturally occurring in the sample) and extrinsic (added to the sample). Intrinsic fluorophores include the aromatic amino acids tryptophan, tyrosine, and phenylalanine (see Appendix E) and chlorophyll. Figure 7.21 shows some examples of extrinsic fluorophores. Fluorescein and rhodamine B are commonly used all-purpose probes that can be conjugated to different molecules. Fluorescent dyes can be introduced in different ways. Either they can be conjugated to a molecule similar to those in the system, for example, a lipid or protein by a permanent covalent bond or the dye molecule can be directly mixed/introduced into the system without binding to a molecule and will respond to the surrounding molecular environment by partitioning into a particular phase or molecular environment.

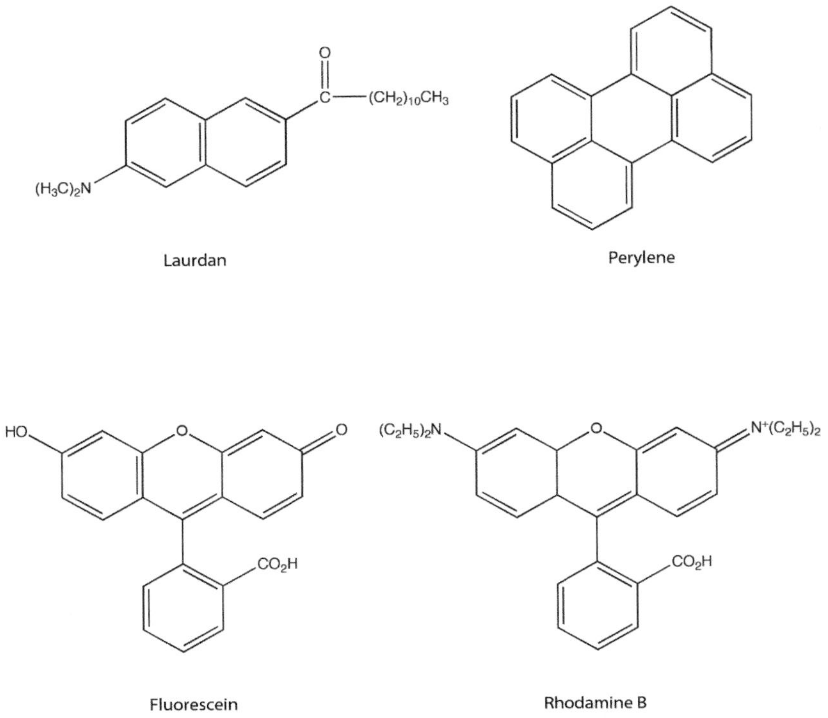

Laurdan

Perylene

Fluorescein

Rhodamine B

FIGURE 7.21 Some examples of common fluorophores used in fluorescence microscopy.

7.5.3 CONFOCAL FLUORESCENCE MICROSCOPY

Many self-assembled biological structures are three dimensional, so how they can be imaged effectively?. A good example of this imaging problem is in the cytoskeletal networks we discussed in Section 7.4.2. A purely 2D image, obtained using the conventional light microscope, does not accurately represent the 3D nature of a filament network. In this case, a powerful tool for imaging is the *confocal microscope*. This microscope is usually used in fluorescence imaging studies, although bright-field images may also be obtained.

Similarly to conventional fluorescence microscopy, materials can be labeled with a fluorescent molecule. When this *fluorophore* is excited at a certain wavelength close to the peak of its absorption spectrum, the emitted light will have a longer peak wavelength and a characteristic emission spectrum (Figure 7.20). The emitted light is collected to produce the image. The excitation wavelength is filtered out before reaching the eyepiece (or camera) to distinguish fluorescence emission (what we want to see) from general reflection of the incident light.

The beauty of the confocal microscope is its ability to capture a 3D image, by taking several images (slices) at different focal depths (along the z axis). These slices can then be combined using software to render a full 3D image or to look at orthogonal cuts through the imaged material. The technique can also be employed to obtain very clear 2D images of materials owing to the excellent z resolution of the microscope. Figure 7.22 shows a confocal microscope image of an F-actin network.

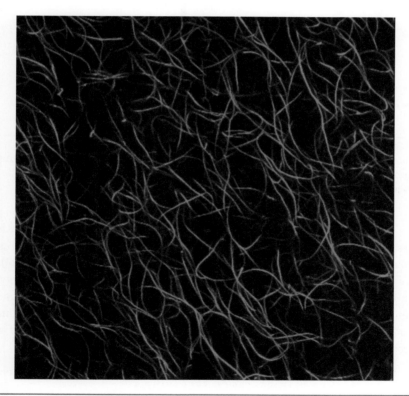

FIGURE 7.22 A fluorescence confocal microscopic image of actin bundles prepared *in vitro*. The bundles form a tangled network of filaments difficult to image using conventional microscopy. To obtain this clear image, a stack of several images at different focal points (*z* positions) was taken with a *z* resolution of about 0.75 μm, then combined. The result is an image with a focal depth of several microns, but with very little out-of-focus light reducing image quality.

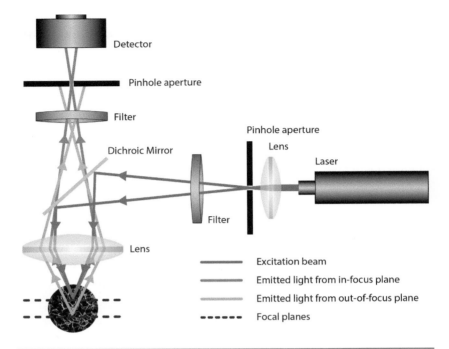

FIGURE 7.23 Diagram demonstrating the working principle of the confocal microscope. Two pinholes act to select light emitted from a narrow focal plane in the same, providing a sharp image with low background noise. These thin "slices" can be collected in sequence to build up a three-dimensional picture of the material.

This ability to image well-defined focal depths in the sample (with greatly improved z resolution over a standard fluorescence microscope) is made possible by the use of two pinholes in the microscope (Figure 7.23). As you can see in the figure, these pinholes effectively cut out all light not emitted from the focal plane, producing a sharp z resolution. In a conventional microscope, these pinholes are not present, so some light from out-of-focus planes can also be seen in the image; this degrades the z resolution significantly.

7.5.4 OTHER FLUORESCENCE TECHNIQUES

In addition to facilitating detailed imaging of 3D structures within a material or biological cell, fluorescence imaging can be adapted to a variety of specialized techniques for biological imaging. The ultimate goal in many applications is to obtain single-molecule information from optical microscopic techniques despite the intrinsic limits of optical resolution.

In FRET (Forster Resonance Energy Transfer) experiments, two different molecules in the sample are tagged with complementary fluorescent labels—one with an emission spectrum overlapping the absorption spectrum of the other. The goal of the experiment is to determine if two different molecules are colocalized (i.e., very close to each other) and therefore potentially associated. If the two molecules are close enough to each other (less than ~ 8 nm), then the emitted light from one fluorophore is likely to be reabsorbed by the second. Detection of a strong emission spectrum specific to the second fluorophore indicates that the molecules are close to one another. This technique provides an important complement to simply imaging colocalization of fluorescent species because it allows the experimenter to distinguish between: (a) pairs of molecules that appear to be associated within the limits of optical resolution, but are in fact not in general close enough to be bound and (b) those likely to be bound because they are close enough for fluorescence energy transfer.

Total internal reflection fluorescence (TIRF) microscopy has gained recent popularity in the field of single-molecule biophysics because to allows us to resolve single, labeled molecules with high resolution. Experimenters have notably used the technique to visualize the motion of single molecular motors walking along microtubules in the cell.[7] The TIRF technique uses the phenomena of *total internal reflection* to image molecules just below the surface of a glass coverslip from which incident light is reflected. When light is totally internally reflected from the upper surface of the coverslip, a small, exponentially decaying amount of light does pass through the glass into the sample. This is known as the *evanescent wave.* This evanescent light will excite fluorophores in the sample within a thin layer close to the interface, although light does not propagate into the bulk (reducing background fluorescence significantly). This technique requires that the molecules of interest are confined to a thin plane very close to the interface.

The FRAP (fluorescence recovery after photobleaching) technique uses photobleaching to measure molecular diffusion and can be used if the material in question is confined to a specific plane (e.g., a membrane or cytoskeletal filaments adsorbed on a surface). The fluorophores in a small area of the sample can be deliberately bleached (i.e., exposed to a high-intensity light for several seconds), then using a lower light intensity, the fluorescence recovery of the area is tracked as unbleached molecules from the surrounding regions diffuse into the bleached area and bleached molecules diffuse out. This experiment can be used to calculate diffusion constants for labeled molecules or those associated with the fluorescently labeled molecules.

7.5.5   TRANSMISSION ELECTRON MICROSCOPY ON SOFT BIOLOGICAL STRUCTURES

Transmission electron microscopy is an imaging technique that uses differences in transmission of electrons through a thin layer of material to produce an image. This technique is widely used in the biological sciences as well as increasingly in materials science. The beauty of using electron microscopy is that the technique can be employed to image structures much smaller than optical techniques will allow (i.e., below the optical diffraction limit).

To create an image, an electron beam is focused onto the sample using an array of beam-steering magnets. As the electrons are focused and pass through the sample, selective absorption occurs because different materials absorb more electrons than others. Staining can be used to enhance this effect. For example, the electron-dense stain uranyl acetate is used to label proteins and is extremely effective in enhancing the contrast of a sample. An example of TEM imaging is shown in Figure 7.24 for bundles of actin filaments prepared using this technique.

To obtain a TEM image, the biological materials must be observed in a vacuum, and this does put some constraints on the range of samples that can be prepared. Any biological sample must be dried, and this is a problem if we wish to represent the 3D structure of the material in solution accurately. A great deal of interesting information on nanoscale structures can still be obtained from dried samples using TEM, however, and the technique is widely used in characterization studies of protein assemblies, membrane structures, and other soft materials, including block copolymer phases and colloidal crystals.

7.5.6   X-RAY SCATTERING FROM BIOLOGICAL ASSEMBLIES

Soft biological samples such as hydrated lipid phases or proteins in solution can be structurally characterized using x-ray diffraction. By treating biological solutions and phases as soft materials, we can use the same analysis techniques applied to other soft materials to investigate biomolecular assembly.

X-ray scattering experiments on biological materials can be divided into two primary techniques, diffraction and scattering. In previous chapters, we discussed the results of diffraction from the "relatively" well-ordered liquid crystal materials and surfactant phases. Many biological materials will scatter x-rays strongly, and they fall into this class of materials. With the correct sample preparation, Bragg scattering can be observed from structures such as the lamellar or hexagonal phases of lipids in water or self-assembled protein structures. A simple theory for Bragg diffraction was discussed in Chapter 3 and scattering in Chapter 5 for polymer solutions (Section 5.8.1).

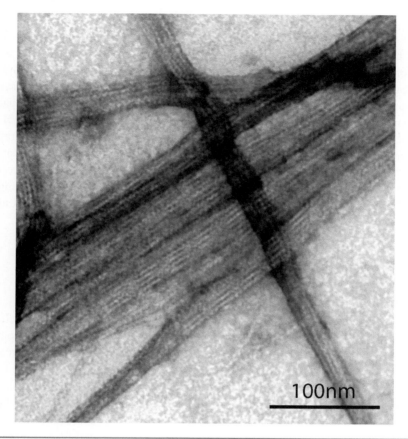

FIGURE 7.24   Transmission electron microscopic image of F-actin filaments stained with uranyl acetate and packed closely together into bundles.

For a more theoretical approach to x-ray scattering, look at Appendix C; I also recommend the excellent book *Modern X-ray Physics* by Als-Nielsen and McMorrow[8] for clear explanations of modern x-ray techniques.

Here, we can discuss x-ray scattering and diffraction results for two different biological structures, demonstrating how x-rays can be used to characterize soft biological structures. In Chapter 2, we introduced Bragg scattering, a technique analogous to the multiple-slit experiment, in which diffraction from a grating results produces a series of diffraction peaks, decreasing in intensity with diffraction order (n). In a similar way, scattering from unordered structures can be characterized by comparing the results to the single-slit experiment. A sample composed of a dilute suspension of globular protein molecules (roughly spherical in shape) will result in a diffraction pattern similar to that from a circular aperture.

The x-ray techniques commonly used for soft biological materials can be classified in different ways. In diffraction-based studies, the primary goal is to observe Bragg scattering from a lattice in the material. Bragg diffraction peaks result from constructive interference between scattered x-rays. Scattering studies are concerned with an analysis of the form factor (see Appendix C) of individual particles in the material. It should be noted that many experiments in biological systems produce results that are a combination of both of these, and lattice d-spacings are large (nanoscale). X-ray scattering experiments can be generally classified as "small-angle" x-ray scattering (SAXS); large-scale structures in the material are probed with scattering vectors up to a q value of about 3 nm^{-1} or "wide-angle" x-ray scattering (WAXS); small-scale structures down to atomic scales can be detected at scattering vectors from about 3 to 30 nm^{-1}.

7.5.7   EXAMPLES OF X-RAY SCATTERING ON SOFT BIOLOGICAL STRUCTURES

7.5.7.1   Lipid Bilayers

Lipids are surfactants, and as we discussed in Chapter 4, they can self-assemble into a variety of lyotropic phases in aqueous solution. Each lipid phase has a distinct structure and a characteristic diffraction pattern detectable by SAXS and WAXS techniques, so x-ray techniques provide a reliable method for phase identification. In addition to the different lyotropic phases, we can also distinguish between the thermotropic variants of the lamellar phase (L_α and $L_{\beta'}$; see Section 7.3.1) using X-ray diffraction. In the L_α phase, the membranes have short-range (liquid crystal-like) in-plane ordering between hydrophilic head groups, whereas the $L_{\beta'}$ (gel) phase has a long-range orthorhombic structure. In the gel phase, a diffraction peak at high scattering angles ($q \sim 15$ nm^{-1}) comes from Bragg diffraction from the ordered lipid head groups. This peak is not present in the L_α phase due to the lack of long-range in-plane ordering.

Figure 7.25 shows an example of SAXS data comparing the gel and liquid crystalline lipid phases. The difference between the two phases is immediately apparent. First, in the gel phase we observe several orders of Bragg diffraction, whereas the liquid crystalline phase typically produces a maximum of two (indicative of liquid crystalline-like short-range ordering). Second, the diffraction peaks observed in the gel phase are narrower than those observed in the liquid crystalline phase. These peaks originate from the regular d-spacing between bilayers. The gel phase is stiffer than the liquid crystalline phase and therefore exhibits fewer membrane fluctuations. Fluctuations will, on average, increase the variation in bilayer spacing throughout the sample, therefore broadening the diffraction peak plotted here.

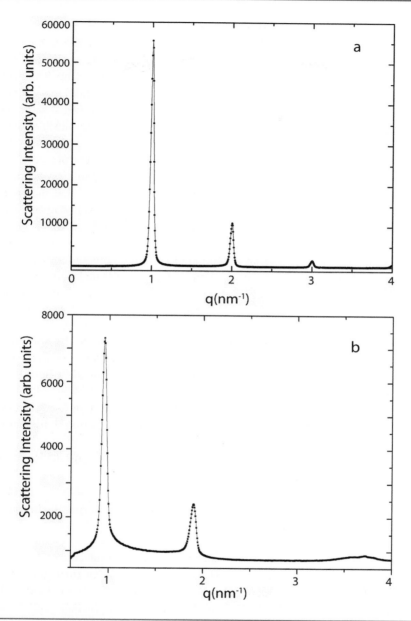

FIGURE 7.25 Examples of SAXS data collected in two different lipid lamellar phases. (a) The gel phase $L_{\beta'}$ on a sample of pure DPPC (1,2-dipalmitoyl-*sn*-glycero-3-phosphocholine) and (b) in the liquid crystalline L_α lipid phase on a sample of pure DOPE (1,2-dioleoyl-sn-glycero-3-phosphoethanolamine). Both sets of data were collected using a synchrotron x-ray source at Brookhaven National Laboratory.

7.5.7.2   Protein Filaments

Here, we can use the example of the actin filaments and bundles to demonstrate the analysis of SAXS solution scattering. In Section 7.4.1, we introduced the filamentous protein F-actin and its self-assembly. The filaments will form networks or bundles of filaments under different solution conditions; we can use fluorescence microscopy to observe these structures in the cell. F-actin is a negatively charged filament, and we can also think of it as a semi-flexible, long, rod-like colloid when in solution.

In an experiment designed to investigate the structures present in a solution of F-actin filaments, we can perform a transmission x-ray scattering experiment. The geometry of this experiment is similar to that shown in Figure 7.26. In the case of dilute solutions such as this, a synchrotron x-ray source is essential because the total scattering intensity from this type of material is very weak.

Ideal x-ray scattering from a solution of F-actin filaments should reveal the structural characteristics of the system from length scales up to about 100 nm and down to the atomic level. Examples of information that may be obtained from unaligned solution scattering include:

✦ Bragg diffraction from any clusters of filaments in the solutions with a lattice-like packing arrangement (e.g., lattice constants corresponding to the 2D arrangement of filaments inside a bundle of filaments)

✦ Correlations between protein molecules in the system (e.g., if there is consistent average spacing between cross-linkers on the filament, this may be detected)

✦ The size and geometry of the assembly or any other consistently sized particles in the solution

✦ The helical structure of F-actin itself. This information is not ideally obtained in an unaligned sample; aligned scattering techniques would be preferred in this case.

Any regular structures (i.e., lattice spacings, molecule-molecule correlation lengths, similar-size particles, etc.) that are globally present throughout the sample can result in a characteristic scattering signature. Theoretically, all of the possible structural information described in the previous sections is available to the experimenter from a single system, but in reality, scattering from a particular structural feature may be too weak to observe. The intensity of a diffraction or small-angle scattering feature depends on several factors, including the concentration of the sample, the degree of ordering for a particular structure (long range/short range, i.e., multiple Bragg orders, may not be present), and the spread in size distribution of a particular feature

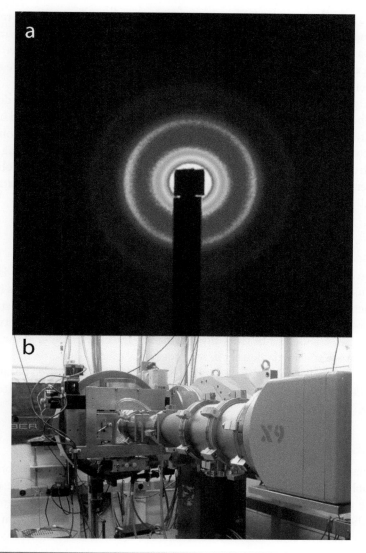

FIGURE 7.26 In a SAXS experiment, scattered x-rays may be collected on a CCD (charge-coupled device) area detector. Unaligned solution samples form a circularly symmetrical intensity pattern (shown here in [a] as light rings on a dark background). The intensity of these rings can then be plotted as a function of the scattering vector q for some appropriate range. Here, we can see an example of an image collected on the CCD detector on Beamline X6B at the National Synchrotron Light Source at Brookhaven National Laboratory in the United States (b), with the detector in the foreground. The sample under investigation was the counterion-induced actin filament bundles also shown in Figure 6.23.

(e.g., a range of bundle thicknesses will broaden the x-ray signature representing bundle thickness). In a real experiment, the scattering vector q-range and parameters for the beam-line must be carefully designed to achieve specific experimental goals, and experimental configuration can also have a big impact on the results.

Figure 7.26 shows an example of SAXS data collected on an area detector. The material under investigation was a solution of F-actin, bundled using $MgCl_2$ (you can see the same system imaged in Figure 7.24). Rings on the detector indicate the packing of filaments in the bundle. The scattering pattern from an unaligned solution sample has a circular distribution. To obtain scattering intensity I graphs as a function of scattering angle θ, the data are radially averaged, then plotted as a function of scattering vector q,

$$q = \frac{4\pi}{n\lambda} \sin\theta \qquad (7.5)$$

where λ is the incident x-ray wavelength. This method of plotting the data is very useful because it provides a way to look at x-ray scattering results independent of experimental parameters, such as sample-to-detector distance and x-ray beam wavelength. These parameters can vary greatly between experiments and must be carefully recorded by the experimenter at the time the data are collected. Plotting in reciprocal space (i.e., using q on the x axis) allows data collected at different sources to be directly compared because q is calculated from the experimental parameters, wavelength, and scattering angle.

7.5.8   NUCLEAR MAGNETIC RESONANCE IN BIOLOGY

In Chapter 5 (Section 5.8.2.3), we introduced NMR as a characterization tool for polymers. NMR is sensitive to molecular conformation, and the technique can be used to identify atoms present, different functional groups, and their configurations. Proteins are biological polymers folded from long chains of amino acids, and their functions are determined by both structure and conformational dynamics. In structural biology, NMR can be a powerful tool for investigating protein function by providing information on both structure and dynamics, including folding and unfolding, chemical kinetics, and rotational and diffusive motion.

In solution-state NMR, the molecules under investigation are free to tumble, averaging out any anisotropic chemical shifts. This technique results in sharp, well-defined peaks; an example of some data can be seen in Figure 7.27. Many proteins, however, cannot be solubilized, and this presents a problem. For example, materials such as self-assembling protein fibrils (e.g., amyloid fibrils) or membrane proteins cannot be solubilized and must be studied *in situ*.

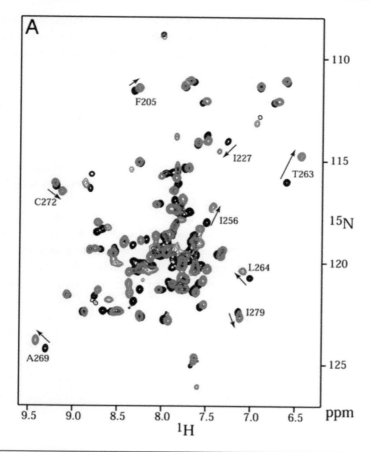

FIGURE 7.27 NMR measurements on protein complexes can reveal changes in structure as the proteins bind to form functional complexes. In this figure, the NMR spectra of free (black) and complexed (red) KaiC proteins are shown to vary. Some particularly notable changes in the chemical shift are marked with arrows. This kind of NMR spectrum (heteronuclear single-quantum coherence) uses two different spin labels (^{1}H and ^{15}N) to map changes in the protein structure. (From Vakonakis, I. and Li Wang, A.C., *Proc. Natl. Acad. Sci. U.S.A.*, 101, 10925–10930, 2004. With permission.)

It is also not possible to crystallize such proteins, thus precluding the use of x-ray crystallography. In this case, solid-state NMR (ssNMR) is a useful technique. A static ssNMR experiment would produce broader peaks than a solution-based experiment due to the lack of molecular tumbling; however, there is a way to get around this problem, and most ssNMR experiments are carried out using magic angle spinning (MAS). In this technique, the sample is rotated rapidly about an axis at a fixed angle to the applied magnetic field.

If the spin frequency is high enough, then sharp peaks comparable to those in solution NMR can be achieved. In soft biological materials, both solution and ssNMR can provide a valuable complement to structural characterization using x-ray diffraction.

QUESTIONS

Biomaterials as Soft Matter

1. Explain why lipids will self-assemble into a bilayer structure. In a non-polar solvent such as chloroform, do you expect to see bilayer formation? How will different solvents affect the assembly?

2. Polyunsaturated lipids have long, flexible alkyl chains, often with several unsaturated bonds. How will this chain structure affect the average space filled by the molecule compared with a saturated lipid? What potential effects may result from adding polyunsaturated lipids to the lipid bilayer if the bilayer leaflets can have different lipid compositions?

3. Estimate the Bjerrum length for the actin filament. How does this length scale compare to the Debye screening length in a 20 mM $MgCl_2$ solution?

4. The persistence length of the actin filament has been experimentally measured to be approximately 10 μm. From this value, estimate the bending energy k_b of the filament. How does this compare to other biological filaments such as DNA or microtubules?

Experimental Techniques

5. X-ray scattering data are collected for a lamellar lipid phase in aqueous solution, and Bragg peaks are detected at $q = 1.04$ and 2.08 nm^{-1}
    a. Calculate the average bilayer spacing in this lipid phase and compare your answer with the expected bilayer thickness based on an approximation for the lipid DOPC (1,2-dioleoyl-sn-glycero-3-phosphocholine). What fraction of the phase is water?
    b. In the gel phase, a WAXS peak is detected at 14.8 nm^{-1}; use this value to estimate the average area taken up by each lipid head group.

6. The maximum resolution for a standard light microscope is given by the formula

$$d = \frac{0.61\lambda}{n\sin\theta}$$

For optical wavelengths, estimate a lower limit for the resolution of an index-matched oil immersion objective. What electron energy would be comparable with this resolution?

7. In TIRF microscopy, the intensity of the evanescent wave in the sample decays as:

$$I = I_0 e^{-z/d},$$

where z represents the distance into the sample, and d is a characteristic penetration depth. If the intensity falls by $1/e$ in 200 nm, calculate the reduction in intensity 1 μm below the surface.

REFERENCES

1. E. Gorter and F. Grendel, On biomolecular layers of lipids on the chromatocytes of the blood. *J. Exp. Med.* 41, 439–443 (1925).
2. P. Yaeger (Ed.), *The Structure of Biological Membranes.* Boca Raton, FL: CRC Press (2004).
3. S.J. Singer and G.L. Nicolson, The fluid mosaic model of the structure of cell membranes. *Science* 175, 720 (1972).
4. K. Simons and E. Ikonen, Functional rafts in cell membranes. *Nature* 387, 569 (1997).
5. T. Sanchez, D.T.N. Chen, S.J. DeCamp, M. Heymann and Z. Dogic, Spontaneous motion in hierarchically assembled active matter. *Nature* 491, 431–434 (2012).
6. E. Sackman, Supported membranes: Scientific and practical applications. *Science* 271, 43–48 (1996).
7. Y. Harada and T. Yanagida, Direct observation of single kinesin molecules moving along microtubules. *Nature* 380, 451–453 (1996).
8. J. Als-Nielsen and D. McMorrow, *Modern X-ray Physics*, 2nd ed. New York: Wiley (2011).

FURTHER READING

D.H. Boal, *Mechanics of the Cell.* Cambridge, UK: Cambridge University Press (2002).
J.R. Lakowicz, *Principles of Fluorescence Spectroscopy*, 3rd ed. New York: Springer (2006).
P. Nelson, *Biological Physics, Energy Information and Life.* Houston, TX: Freeman (2004).
W.C.K. Poon and D. Andelman (Eds.), *Soft Condensed Matter Physics in Molecular and Cell Biology (Scottish Graduate Series).* Boca Raton, FL: CRC Press (2006).
N. Stribeck, *X-ray Scattering of Soft Matter.* New York: Springer Laboratory (2007).
R.G. Weiss and P. Terech (Eds.), *Molecular Gels: Materials with Self-Assembled Fibrillar Networks.* New York: Springer (2006).

Glossary

active matter: an out-of-equilibrium system in which the constituent particles consume energy from the environment and convert this energy into work. Active materials often exhibit large-scale internal dynamics and collective motion.

alignment layer: a coating on the inner surfaces of a liquid crystal device designed to produce a uniformly oriented liquid crystal domain.

anisotropy: a property of a material indicating directional non-uniformity. Anisotropic materials exhibit a property that varies depending on the direction of measurement. In an optically anisotropic material, more than one refractive index can be measured.

biomechanics: the study of the mechanical properties of macroscopic biological systems.

biomolecular assembly: the spontaneous packing of biological molecules into larger-scale structures resulting from the action of intermolecular forces.

biopolymer: a polymer-like filamentous biological molecule composed of a large number of repeat units.

birefringence: a property of optically anisotropic materials in which two or more different orientationally dependent refractive indices are present.

block copolymer: a polymer composed of two or more distinct blocks of repeating subunits.

Boyle temperature: the temperature at which a non-ideal gas will exhibit the properties of an ideal gas.

Brownian motion: random directional motions of a particle or molecule resulting from collisions with the molecules that comprise the medium in which the particle is suspended.

calamitic liquid crystal: a liquid crystal phase consisting of rod-shaped molecules or particles.

chemical shift: the relative shift of a nuclear magnetic resonance signal relative to a reference point as a result of the local chemical environment.

chiral: chiral molecules have a "handedness," an inherent twist in their structure, and do not pack directly on top of each other. A chiral structure cannot be superimposed on its mirror image.

cholesteric phase: a liquid crystal phase characterized by nematic-like ordering with a helical twist.

Clausius-Clapeyron equation: this equation describes the shape of the phase boundary on the phase diagram.

cluster–cluster aggregation: a flocculation mechanism in colloidal systems characterized by the successive aggregation of larger and larger clusters in solution.

coarsening: *see* Ostwald ripening

colloidal crystal: a close-packed structure formed from colloidal particles analogous to an atomic crystal.

confocal microscope: an optical microscope designed using pinholes to achieve very good z resolution.

conformation: the exact shape of a molecule at a moment in time.

copolymer: a polymer composed of more than one repeating subunit.

correlation length: a characteristic length between particles in a material.

cream: the process by which colloidal particles less dense than the continuous medium in which they are suspended float to the surface and separate out.

creep: time-dependent permanent deformation of a material under the application of a constant applied stress.

critical micelle concentration: the concentration of surfactant in a solvent above which micelle formation occurs.

critical point: a unique point on the phase diagram at the end of a phase boundary line where the two phases become indistinguishable.

cytoskeleton: the filamentous network of proteins in the cell consisting of F-actin, microtubules, and intermediate filaments.

Debye screening length: characteristic length scale for charged colloidal particles in an ionic solution beyond which the Coulomb force is effectively screened.

defect texture: characteristic birefringence pattern observed on the polarizing light microscope for a liquid crystal material.

dielectric anisotropy: the difference in dielectric constants for light parallel and perpendicular to the molecular director in a liquid crystal material.

director: in liquid crystal science, the director **c** is a vector that represents the local average molecular orientation in a liquid crystal phase.

discotic liquid crystal: a liquid crystal phase formed from disk-shaped molecules or particles.

disclination: a topological defect line in a material's liquid crystalline structure characterized by a discrete change in molecular orientation.

dispersity: *see* polydispersity

elastic: elastic materials extend reversibly under an applied force; they typically obey Hooke's law under small deformations.

electric double layer: the distribution of charges around a colloid suspended in an ionic solution.

electrophoresis: the migration of charged particles in solution in an electric field.

emulsion: a suspension of droplets of one normally immiscible fluid phase in another stabilized by a surfactant.

enthalpy: the total energy it takes to create a material including effects on the environment. One of the thermodynamic potentials.

ferroelectricity: the response of a material with a spontaneous polarization to an applied electric field. Ferroelectrics exhibit a hysteretic behavior and can be reoriented in response to a changing electric field direction.

first-order phase transition: a thermodynamic phase transition at which the order parameter is discontinuous.

floc: a large aggregate of weakly bound colloidal particles.

flocculation: reversible aggregation of colloidal particles into macroscopic clusters.

fluid mosaic model: a model for the structure of the cell membrane in which the membrane constituents do not exhibit any in-plane organization.

fluorescence: emission of light from a material after excitation and subsequent relaxation of an electron from a singlet state.

form factor: a function describing the scattered wave pattern resulting from the electron density distribution in a material.

Fourier transform: a mathematical operation by which a function can be expressed in terms of a sum of sine and cosine functions.

frank elastic constants: constants describing the elastic response of a liquid crystal material under the splay, twist, and bend deformations.

Fréedericksz transition: induced alignment of liquid crystal molecules with the field lines of an applied electric or magnetic field.

free energy: a thermodynamic potential (either the Gibbs free energy or the Helmholtz free energy).

freely jointed chain model: *see* ideal chain model

gel: a viscoelastic state of matter typically with high solvent content.

Gibbs-Marangoni effect: a mass transfer phenomenon observed in fluids with a surface tension gradient.

glass transition: a change of state observed in many materials. On cooling, the material forms a glassy state with a very long relaxation time and amorphous structure.

hard sphere repulsion: a strong repulsive potential representative of the inability of atoms or particles to both occupy the same space deriving from the Pauli exclusion principle.

heat of transformation: *see* latent heat

homeotropic alignment: liquid crystal molecules are aligned in such a way that their long axis lies perpendicular to a substrate.

homopolymer: a polymer formed from a single repeated monomeric unit.

hydrogel: a gel formed with water as the continuous-phase component.

hydrogen bonds: low energy, often temporary, bonds formed between polar molecules.

hydrophilic: a material or substance that tends to locate in water.

hydrophobic: a material or substance that tends to avoid water.

ideal chain model: in polymer science, a model for the polymer chain in which the monomers are freely joined with no interactions between monomers.

ideal gas model: a model describing the behavior of a dilute gas. Intermolecular interactions are excluded from the model.

in-plane device: a liquid crystal device in which an electric field is applied parallel to the plane of the device.

isoelectric point: the pH at which a molecule has an equal number of positive and negative charges in solution (i.e., it has no net charge).

isotropic phase: a thermodynamic phase of a material in which the material properties are not dependent on sample orientation.

jamming: a phenomenon in granular materials in which the material will suddenly cease to flow or clog as a result of interparticle interactions.

Langmuir-Blodgett film: a monolayer surfactant film on a fluid surface typically formed using a Langmuir-Blodgett trough.

latent heat: the change in enthalpy associated with a first-order phase transition.

lattice factor: function describing the effects of the crystal lattice structure on scattered x-ray wave intensity.

Lennard-Jones potential: a model used to describe the potential between neutrally charged particles or atoms, including a hard-sphere-like repulsion at very short length scales and a van der Waals attraction at longer length scales.

lipid bilayer: a double layer of lipid molecules that self-assembles in aqueous solution with hydrophilic head groups facing out toward the water.

luminescence: the emission of light from a material resulting from an electronic excitation.

lyotropic liquid crystal: a material that exhibits different liquid crystal phases as a function of concentration in a solvent.

macroscopic: structure or properties on a large length scale where the states of individual molecules are not considered.

mean free path: the average distance over which a molecule or particles will move before interacting with another particle.

micelle: a small aggregate of surfactant molecules in solution in which the hydrophobic tails are oriented inward away from the water.

microrheology: a rheological technique in which small particles are dispersed throughout a material and their motions tracked.

microscopic: having a size of the order of micrometers or less.

Mie scattering: light scattering from particles where the particle size is similar to the incident wavelength.

model membranes: lipid bilayers composed of purified lipid components and designed to be used as a simplified experimental model for the cell membrane.

Newtonian fluid: a gas or liquid for which the viscosity does not depend on shear rate at a fixed temperature and pressure. Water is a Newtonian fluid.

Ostwald ripening: a mechanism by which larger domains grow. Molecules from smaller domains diffuse into larger domains, favoring the growth of larger regions.

packing parameter: a parameter used to quantify the effects of surfactant geometry on molecular packing.

Pauli exclusion principle: the idea that no two electrons in an atom can be in the same quantum state at the same time.

percolation: the point at which a system aggregates and becomes connected such that a path can be defined that can be followed from one side of the system to the other.

persistence length: a measure of filament flexibility representing the distance over which a polymer chain loses orientational information.

phase separation: the spontaneous separation of different molecular species when mixed due to unfavorable intermolecular interactions.

phonon: a quasi-particle representing a quantized lattice vibration in a crystal.

phosphorescence: light emitted from a material after excitation of an electron to a triplet state and subsequent relaxation.

photonic crystal: a material with a regular crystal-like structure and a lattice spacing on the order of visible light. Such structures can be used as waveguides.

planar alignment: liquid crystal molecules are aligned in such a way that their long axis lies parallel with a substrate.

Poisson's ratio: the ratio between the transverse strain on a material perpendicular to the direction of the force applied and the linear strain in the direction of the force.

polarized optical microscopy: a microscopic technique in which the sample is viewed in transmission in-between crossed polarizers.

polydisperse: *see* polydispersity

polydispersity: the distribution of molecular lengths found in a polymer material.

polysaccharide: a polymer composed of sugar monomers such as glucose.

principle curvature: the deviation of a surface from a flat plane as defined by the radii of curvature in two perpendicular directions.

radius of gyration: a measure of the space occupied by a polymer chain in solution.

random walk: the random spatial path taken by a particle experiencing Brownian motion.

Rayleigh scattering: light scattering from particles where the particle size is much smaller than the incident wavelength.

relaxation time: the time it takes for a material to return to equilibrium after a deformation.

rheology: the study of the viscoelastic properties of materials.

rheometer: an experimental device for measuring the viscoelastic properties of a material.

ripple phase: a thermotropic phase observed in lipid bilayers in which the membrane exhibits an out-of-plane rippled configuration.

rubber: a polymer melt with elastic properties.

Schlieren texture: a planar liquid crystal birefringence texture characterized by dark "brushes."

second-order phase transition: a thermodynamic phase transition characterized by continuous variation of an order parameter across the transition.

sediment: the settling of colloidal particles out of suspension over time.

self-similar: a material characterized by a structural characteristic that is repeated on different length scales. These structures are fractal and can be characterized by a fractal dimension.

semi-flexible polymer: a polymer chain with a high bending rigidity in between that of a Gaussian chain and a rigid rod.

shear stress: the ratio of the applied force on a material to the cross-sectional area parallel to the direction of force application.

shear thickening: a phenomenon in which the viscosity of a material increases as shear rate is increased.

shear thinning: a phenomenon in which the viscosity of a material decreases as the shear rate is increased.

smectic: a liquid crystal phase in which the molecules are arranged in a layered structure.

smectic layer spacing: the distance between consecutive layers of molecules in a smectic liquid crystal phase.

sol: a dispersion of colloidal particles with fluid-like (viscous) properties.

steric: a term referring to the shape of a molecule or particle, that is, the arrangement of its constituent atoms in space.

Stern layer: a highly charged layer of condensed ions on the surface of a colloidal particle in ionic solution.

strain: a dimensionless number representing the ratio of elongation of a material under stress with respect to the original length.

stress: the ratio of the applied force on a material to cross-sectional area perpendicular to the direction of force application.

stress relaxation: the time-dependent decrease in stress in a material held at constant strain.

structure factor: a function that describes the distribution of scattered waves from a single unit cell in a material.

supported bilayer: a lipid or other surfactant bilayer deposited on a non-rigid substrate such as a polymer network.

thermal equilibrium: a state in which a two- or more-body system does not experience heat flow between those bodies.

thermotropic liquid crystal: a material that exhibits different liquid crystal phases as a function of temperature.

theta solvent: a solvent in which a polymer molecule behaves as an ideal Gaussian chain.

theta temperature: the temperature at which a polymer in solution behaves like an ideal Gaussian chain.

thixotropic: a shear-thinning material that exhibits a time-dependent recovery in viscosity after removal of the shearing force.

topological defect: a point in a liquid crystal system at which local ordering is undefined.

triple point: the point on a pressure/volume/temperature (PVT) phase diagram at which a substance can exhibit the solid, liquid, or gas phases.

Tyndall effect: visualization of colloidal particles in a fluid in the path of a light beam by light-scattering effects.

van der Waals potential: an attractive potential between neutrally charged atoms or particles.

vertical alignment display: a liquid crystal display in which the liquid crystal molecules are aligned perpendicular to the plane of the screen (homeotropic alignment).

viscoelastic: viscoelastic materials exhibit both viscous and elastic behaviors depending on the timescale over which a deformation is applied.

viscosity: a bulk material property that characterizes a fluid's resistance to flow.

vulcanization: a process used in rubber manufacture in which a cross-linking agent renders the rubber more durable and elastic.

zeta potential: the potential difference between the edge of the stern layer and the diffuse layer around a colloidal particle in an ionic solution.

Zeeman splitting: splitting of the electronic spin states in a material induced by an applied magnetic field.

zwitterionic: a zwitterionic molecule has both positive and negative charges spatially separated from each other.

Appendix A:
The Fourier Transform

The Fourier transform is an important mathematical concept in many different areas of science, including optics, scattering and diffraction measurements, and signal processing, all of which can be applied to the various aspects of soft matter science. The key idea behind the Fourier transform is that any arbitrary function can be represented by a sum of sines and cosines of different frequencies (ω).

By taking the Fourier transform of any function, we can obtain information on all of the different frequencies that make up that function and their relative importance. The Fourier transform allows us to pick out the dominant frequencies or frequency patterns in data that may reveal important information about a material experimentally. Either in the time domain (e.g., the correlation function recorded in dynamic light scattering) or spatially (such as an x-ray diffraction pattern).

It is convenient to represent a sum of sines and cosines as an exponential function, and if $g(t)$ in the following equation is a function in the time domain, then $f(\omega)$ is a function that represents the Fourier transform of $g(t)$.

$$g(t) = \frac{1}{2\pi} \int_{-\infty}^{\infty} f(\omega) \, e^{i\omega t} d\omega,$$

where $f(\omega)$ is in the frequency domain. Alternatively, we can also write the inverse relationship,

$$f(\omega) = \int_{-\infty}^{\infty} g(t)e^{-i\omega t}dt.$$

These equations also apply in space instead of time (or for any pair of inversely related variables). If I define a function $g(x)$, then the Fourier transform of this function $f(k)$ is a function defined by the vector k, where k is a position vector in what is known as "reciprocal space," an idea analogous to frequency. In x-ray or light-scattering experiments, the pattern we observe on the detector is the Fourier transform of the real structure of the material (minus some phase information); it reveals the density distribution of the material.

Appendix B: Physical Constants and Conversions

Speed of light in vacuum	c	$= 2.998 \times 10^8$ m/s
Electron charge	e	$= 1.602 \times 10^{-19}$ C
Electron mass	m_e	$= 9.109 \times 10^{-31}$ kg
Permittivity of free space	ε_0	$= 8.9 \times 10^{-12}$ C^2/Nm^2
Permeability of free space	μ_0	$= 1.26 \times 10^{-6}$ Tm/A
Avogadro's number	N_A	$= 6.022 \times 10^{23}$ mole^{-1}
Boltzmann's constant	k_B	$= 1.381 \times 10^{-23}$ J/K
Stefan-Boltzmann constant	σ	$= 5.7 \times 10^{-8}$ W/m^{2}K^4
Molar gas constant	R	$= 8.314$ J/mol K
Planck's constant	h	$= 6.626 \times 10^{34}$ Js
Acceleration due to gravity on earth	g	$= 9.807$ m/s^2

1 eV $= 1.602 \times 10^{-19}$ J
1 calorie $= 4.184$ J
Temperature (K) $=$ Temperature ($^\circ$C) + 273.15 K
Temperature ($^\circ$F) $= 9/5$(Temperature ($^\circ$C)) + 32
1 Å $= 10^{-10}$ m
1 L $= 10^{-3}$ m^3
1 atm $= 1.013 \times 10^5$ Pa $= 760$ mmHg
A 1-molar solution (1M) contains 6.022×10^{26} molecules/m^3

Appendix C: Laue Scattering Theory

In 1914, Professor Max von Laue was awarded the Nobel Prize in Physics for his discovery that x-rays could be diffracted by crystals. von Laue used a simple transmission x-ray experiment and a copper sulfate crystal to collect the diffraction pattern on a photographic plate. von Laue formulated a simple theoretical description for the resulting diffraction patterns formed on the plate and showed not only that x-rays can be used for diffraction, but also that the crystal lattices postulated in the previous century by Bravais could be observed by this method.[1]

If a crystal is illuminated by an x-ray beam of wave vector $\mathbf{k}$ (where $\mathbf{k}$ is equal to $\frac{2\pi}{\lambda}$, and is the wavelength of the beam), then the general amplitude, A of traveling wave incident on an electron in the material will be equal to:

$$A = A_o e^{i(\mathbf{k} \cdot \mathbf{r} - \omega t)},$$

where A_0 is the amplitude of the beam source, and the position of the electron is defined by the position vector $\mathbf{r}$ from the origin as shown in Figure C.1.

If we assume that the electron is located in a particular atom in the unit cell $\mathbf{r_e}$ from the atomic nucleus and that the position of this atomic nucleus is defined by a vector $\mathbf{S}$ from a corner of the unit cell, located $\mathbf{R}$ from the origin, then the position vector of the electron can also be defined by the addition of these three vectors:

$$\mathbf{r} = \mathbf{R} + \mathbf{S} + \mathbf{r_e}.$$

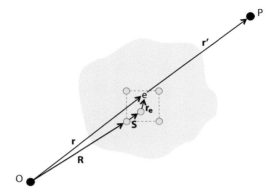

FIGURE C.1 The Laue scattering geometry.

The amplitude of a wave scattered from this electron can therefore be written as:

$$A_s \propto e^{i\mathbf{k}\cdot(\mathbf{R}+\mathbf{S}+\mathbf{r_e})}.$$

From this, we can write an expression for the x-ray amplitude A_s at a point far away from the electron:

$$A_s = Ce^{i\mathbf{k}\cdot(\mathbf{R}+\mathbf{S}+\mathbf{r_e})}e^{i\mathbf{k'}\cdot\mathbf{r'}}\rho(r)d^3\mathbf{r_e},$$

where C is a constant, and $\rho(r)$ represents the electron density at a point $\mathbf{r}$ in the sample; $\mathbf{k'}$ is the scattered wave vector, and $\mathbf{r'}$ is the displacement vector from the electron to the detector.

We can rewrite this equation as:

$$dA_s = Ce^{i\Delta\mathbf{k}\cdot(\mathbf{R}+\mathbf{S}+\mathbf{r_e})+i\mathbf{k'x}}\rho(r)d^3\mathbf{r_e},$$

where $\mathbf{x}$ represents a translation from the origin to the detector.

This equation has been written to represent the general amplitude of scattered x-rays from a single electron in a material; now, we must extend this analysis to the entire crystal if it is going to be useful. We can sum our result over every electron for each atom, then sum for every atom in the unit cell and for every unit cell in the crystal. This gives us the following equation:

$$A = C\left[\sum_R e^{-i\Delta\mathbf{k}\cdot\mathbf{R}}\right]\cdot\left[\sum_S \left(\rho(\mathbf{r_e})e^{-i\Delta\mathbf{k}\cdot\mathbf{r_e}}d^3\mathbf{r_e}\right)e^{-i\Delta\mathbf{k}\cdot\mathbf{s}}\right]\cdot e^{i\mathbf{k'}\cdot\mathbf{x}}.$$

This equation can be simplified if we condense the result into three different factors, each contributing to the overall scattering intensity: the form factor, structure factor, and the lattice factor.

The *form factor* f_0 represents scattering from the electron density distribution in the material.

$$f_0 = \int \rho_s(\mathbf{r_e}) e^{-i\Delta k \cdot \mathbf{r_e}} d^3 \mathbf{r_e}.$$

The *structure factor* S includes this form factor and represents scattering from the entire unit cell. Interference effects between the different atoms in the unit cell can be important in modifying the scattering pattern.

$$S = \sum_S f_0 e^{-i\Delta k \cdot S}.$$

Then, the *lattice factor* L modifies this scattering to include interference effects resulting from the crystal lattice.

$$L = \sum_R e^{-i\Delta k \cdot R}.$$

For a regular lattice structure, the lattice factor modifies the amplitude in such a way that the scattered intensity is small except at certain values. This results in the well-defined intensity peaks characteristic of Bragg scattering.

Altogether, this gives an equation for the scattered intensity as follows:

$$I = |A|^2 = |Ce^{ik' \cdot x} L.S|^2.$$

So what does this give us? When x-rays are scattered from a material, they interact with the electrons in the structure. The scattered waves, and thus the final scattering pattern produced, however, are modified by the arrangement of those electrons on different lengthscales. Materials are hierarchical in their electron density distribution and from the individual atom, to the unit cell, to the crystal lattice, the final scattering pattern will reflect all of this information. Regular structures can produce strong constructive interference (diffraction patterns), whereas amorphous materials produce more diffuse patterns with characteristic features. The difficult part can be interpreting the result!

REFERENCE

1. S. Brandt, *The Harvest of a Century: Discoveries of Modern Physics in 100 Episodes*. Oxford, UK: Oxford University Press, Chapter 20 (2009).

Appendix D: Entropy and Thermodynamic Equilibrium

It is well known in thermodynamics that systems tend towards their equilibrium state. There is an important link between this concept of equilibrium and the quantity entropy. In this appendix, we discuss these concepts and define entropy a bit more specifically. These ideas are not only specific to soft matter, but also can be applied to any thermodynamic system.

Before we start thinking about the macroscopic quantity, entropy S, it is important to think first about ways to describe the number of possible states of a system. For example, a system such as an ideal gas in a sealed container could be: (a) uniform in density throughout the container or (b) all the molecules could be on one side of the container. Both of these states could, in theory happen, but we know which one is the most likely! (b) is so unlikely to spontaneously occur, that in reality we will never observe it.

Thermal physics helps us to link the microscopic description of a material, in which we can specify the state of every particle, to macroscopic measurable parameters. A nice way to start thinking about this concept is to visualize a very simple system. Here, we can use an example of ten coins; each coin can be heads or tails, but the coins are all identical.

If I place all ten coins in a row, tails up, there are ten different ways to flip the coins so they show one head and nine tails. We can describe this by saying that there are ten different ways (microstates) to give the macrostate of "one head." The macrostate "one head" has a multiplicity of 10. If two of the coins are flipped over, "two heads," there are "45 ways" to get this macrostate; a multiplicity of 45. Thinking about the probabilities of turning over different numbers of coins, we can write a general formula so that the multiplicity of a system of N coins with n heads can be calculated using the equation:

$$\Omega(n) = \frac{N!}{n!(N-n)!}.$$

This coin example represents a simple idea to consider as we expand this concept of microstates and macrostates to a much more complex, real system, such as the ideal gas. In a realistic gas, there are about 10^{23} particles and an enormous number of different possible states.

In an ideal gas, the microstate of the system can be described microscopically at any instant in time by knowing the positions and momenta of all of the particles. Even if each particle is identical, and therefore interchangeable, there will be a very large number of possible microstates for each macrostate.

The multiplicity (number of microstates in a particular macrostate) is highly dependent, as you can imagine, on the number of particles, the volume occupied by the gas, and the internal energy of the gas. In an ideal gas, multiplicity can be represented by a general function:

$$\Omega\,(U,N,V) = f(N)V^N U^{3N/2}$$

where $f(N)$ is a function of N, the number of molecules, V is the volume occupied by the system, and U is the internal energy of the system. In a system where N is of the order of Avogadro's number, you can see how just a tiny increase in volume or internal energy in the system will have a vast impact on the number of possible states. In fact, because of the extreme sharpness of this function, the probability of the state with the highest multiplicity occuring, is massively higher than the probability of a state even just a small deviation away. So, once the system reaches the most probable microstate, it is overwhelmingly unlikely that a system will move significantly away from that state. Thus, when a system reaches its most likely microstate, the system can be said to be at equilibrium.

Because of the incredibly large numbers involved when discussing the multiplicity of a real system, one way to express the multiplicity is to change Ω into the more manageable quantity S, the entropy. The quantity entropy can be defined as:

$$S = k_B \ln \Omega,$$

where k_B is the Boltzmann constant. Entropy, as you can see, is just the natural log of multiplicity multiplied by a constant, giving it units of J/K. So, entropy is a macroscopic thermodynamic quantity that represents the number of possible ways the elements of the system can be arranged. At equilibrium, the entropy reaches a maximum value, and once this state is reached, it is so unlikely that the entropy of the system will decrease that we arrive at the second law of thermodynamics:

> *A large system in equilibrium will be found in the macrostate with the greatest entropy.*

This statement is easy to understand now in terms of macrostates and multiplicity. For a large system of 10^{23} particles, probability dictates that the material will be in the most likely macrostate or one very close to this state. Going back to our ideal gas example, say, the gas in a room spontaneously moves to occupy a volume representing 50% of the room's volume. This is a much lower entropy state than for the gas to fill the room uniformly. This scenario is so unlikely that we can state that it will never happen. The overwhelming likelihood that a system will remain in or close to the most probable state (in equilibrium) means that in the thermodynamic limit (for a very large number of molecules, i.e., Avogadro's number), entropy will not decrease spontaneously and can only spontaneously increase.

Appendix E:
The Amino Acids

Chemical structures of the 20 amino acids. Notice that all of the structures are based on the same structure (on the right-hand side of each structure: a central carbon connected to a carboxyl and an amino group), with different side groups.

glycine (GLY) alanine (ALA) cysteine (CYS) serine (SER)

threonine (THR) valine (VAL) isoleucine (ILE) leucine (LEU)

aspartic acid (ASP) asparagine (ASN) glutamic acid (GLU)

glutamine (GLN) methionine lysine (LYS)

arginine (ARG) phenylalanine (PHE) histidine (HIS)

proline (PRO) tyrosine (TYR) tryptophan (TRP)

Index

Note: Page numbers in italic and bold refer to figures and tables, respectively.